POPULATIONS AND ECOSYSTEMS
RESOURCES

IMAGES, DATA, AND READINGS

DEVELOPED AT LAWRENCE HALL OF SCIENCE, UNIVERSITY OF CALIFORNIA AT BERKELEY
PUBLISHED AND DISTRIBUTED BY DELTA EDUCATION

FOSS Middle School Project Staff and Associates

FOSS Middle School Curriculum Development Team
Linda De Lucchi, Larry Malone, Co-directors; **Dr. Lawrence F. Lowery,** Principal Investigator
Dr. Susan Brady, Teri Dannenberg, Dr. Terry Shaw, Susan Kaschner Jagoda, Curriculum Developers
Dr. Kathy Long, Assessment Coordinator; **Anne Gearhart,** Research Assistant
Carol Sevilla, Graphic Artist; **Rose Craig,** Artist
Alev Burton, Administrative Support; **Mark Warren,** Equipment Manager

ScienceVIEW Multimedia Design Team
Dr. Marco Molinaro, Director; **Rebecca Shapley,** Producer
Leigh Anne McConnaughey, Principal Illustrator
Dan Bluestein, Lead Programmer
Tom McTavish, Programmer; **Roger Vang,** Programmer
Coe Leta Finke, Usability Review

Special Contributors
Marshall Montgomery, Materials Design; **John Quick,** Video Production
Dr. Robert Bohanan, Course Content Consultant
Dr. Gerardo R. Camilo, Adviser; **Adrian Barnett,** Adviser

Photo and Information Assistance
**American Lands Alliance, Arctic National Wildlife Refuge, Bureau of Land Management
Cimarron National Grassland, Channel Islands National Park Service
Columbia University Biosphere 2 Center (Oracle, Arizona), Cornell Lab of Ornithology
Delaware Water Gap National Recreation Area, Environment Canada/Yukon, Everglades National Park
Florida Keys National Marine Sanctuary, El Yunque Caribbean National Forest
Gillette Entomology Club at University of Colorado, Melinda S. LaBranche, Mono Lake Committee, Monongahela
National Forest, MODIS Rapid Response Project (NASA/GSPC), Monterey Bay Aquarium Foundation
Monterey Bay National Marine Sanctuary, National Atmospheric Deposition Program, National Park Service
Nebraska Game and Wildlife Commission, NOAA Atlantic and Meteorological Laboratory
NOAA National Marine Fisheries, NOAA National Undersea Research Program
NOAA National Estuarine Research Reserve, NOAA Photo Library, Patuxent Wildlife Research Center
Saguaro National Park, Saint Louis University, U.S. Department of Agriculture
U.S. Environmental Protection Agency, U.S. Geological Survey, U.S. Fish and Wildlife Service
University of Georgia, University of Nebraska Cooperative Extension, University of Texas at Dallas**

Delta Education FOSS Middle School Team
Bonnie Piotrowski, FOSS Managing Editor; **Mathew Bacon, Carrie DiRienzo, Grant Gardner, Tom Guetling,
Joann Hoy, Dana Koch, Cathrine Monson, John Prescott, Karen Stevens, Rebecca Waites**

National Trial Teachers
Judith Hitchings and **Maria Little,** Borel Middle School, San Mateo, CA
Bernice Hagan, Academy Middle School, Fitchburg, MA
Jennifer Fischer, B.F. Brown Middle School, Fitchburg, MA
Sarah Chapin and **Julie McKinney,** Hudson High School, Hudson, MA
Gayle Dunlap and **Donna Moran,** Walter T. Bergen Middle School, Bloomingdale, NJ
Joan Caroselli and **Shirley Thomas,** J.E. Soehl Middle School, Linden, NJ
John Kuzma, Alex Cladakis, and **Brenda Cahill,** McManus Middle School, Linden, NJ
Brad Edwards and **Patricia Volino,** Rahway Middle School, Rahway, NJ
Gina Garlie, Amboy Middle School, Amboy, WA; **Barbara Carroll,** La Center Middle School, La Center, WA
Virginia Reid and **Monica West,** Thurgood Marshall Middle School, Olympia, WA
Jodi Boe, Candyce Burroughs, and **Kathy Kinnaird,** Washington Middle School, Olympia, WA
Kristina Zulick-Roth, Badger Middle School, West Bend, WI

Lawrence Hall of Science

*FOSS for Middle School Project
Lawrence Hall of Science, University of California
Berkeley, CA 94720 510-642-8941*

...because children learn by doing.®

*Delta Education
P.O. Box 3000 80 Northwest Blvd.
Nashua, NH 03063 1-800-258-1302*

The FOSS Middle School Program was developed in part with the support of the National Science Foundation Grant ESI 9553600. However, any opinions, findings, conclusions, statements, and recommendations expressed herein are those of the authors and do not necessarily reflect the views of the NSF.

Populations and Ecosystems
3 4 5 6 7 8 9 QUE 08 07 06 05 04

542-1448
1-58356-438-1

RESOURCES
Table of Contents

Milkweed Bugs

WHAT ARE MILKWEED BUGS LIKE?

Milkweed bugs are easily recognized as insects. They have the same structures as just about all other insects: six legs, three body parts (head, thorax, and abdomen), and two antennae.

Milkweed bugs are true bugs because they do not have mouths for biting and chewing food—they have a tubelike beak for sucking fluids. The scientific name for such a mouth is a **proboscis.**

In nature, the milkweed bug uses its beak to pierce and suck nutrients from the seeds of the milkweed plant. The ones in your classroom, however, have been bred to feed exclusively on raw, shelled sunflower seeds. These milkweed bugs insert their long beaks into sunflower seeds to suck out the oils and other nutrients.

HOW DO MILKWEED BUGS GROW?

Milkweed bugs start life as tiny eggs. When the eggs hatch, about a week after being laid, they are tiny—not much bigger than the period at the end of this sentence. If you look at a newly hatched milkweed bug under a microscope, you will see that it is indeed a tiny bug, with six legs, three body parts, and two antennae. You will also see that it has a tough outer covering called an **exoskeleton** to protect it. The exoskeleton is not flexible, so the tiny bug cannot grow while the exoskeleton is in place. So...

After a few days the milkweed-bug nymph (immature bugs are called **nymphs**) bursts its exoskeleton and slides out of the old shell. The new exoskeleton is moist and flexible, and the bug pumps itself up, growing to twice its size in a matter of minutes. In a few hours the exoskeleton hardens, and the larger nymph goes about its business (eating and growing).

About a week after the bugs hatch, crispy little transparent exoskeletons with black squiggly legs start to accumulate in the bottom of the habitat. The process of shedding the exoskeleton in order to grow is called **molting.** Just after molting, the bug is creamy yellow with bright red legs and antennae. Within a few hours the body turns darker orange, and the legs and antennae become black again.

The milkweed bug molts five times before it becomes a fully mature adult. With each molt the body shape changes, the bug develops more dark body markings, and wings start to form. Each nymphal stage is referred to as an **instar.** The first instar is the newly hatched baby, and the fifth instar is the one just before adulthood.

This gradual maturing of an insect is called **incomplete metamorphosis.** The bug steadily gets bigger and more complete until the last molt reveals the adult. The process from egg to adult takes 4–8 weeks, depending to a large extent on the temperature. A week or more after reaching adulthood, the bugs will mate, and the female will lay eggs. In a room that is a comfortable temperature for humans, the eggs will hatch in about another week, changing from lemon yellow to tangerine orange as they mature. The bugs that hatch out will be about half males and half females. The life cycle from egg to egg is about 2 months.

HOW DO MILKWEED BUGS MATE?

You can easily observe mating, as the two mating bugs remain attached end-to-end for a long time.

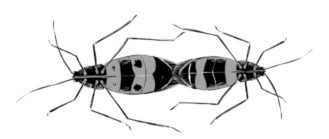

It is possible to distinguish the female adults from the males by the body markings on the ventral (belly) side of the bugs. The tip of

the abdomen is black on both sexes. Next comes a solid orange segment (with tiny black dots at the edges). If the next two segments after the orange segment are solid black bands, it is a male.

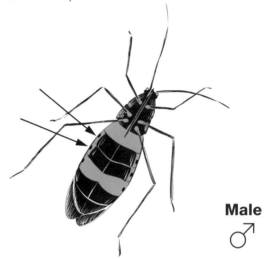

Male
♂

If the next segment after the orange segment is orange with two large black spots, followed by a solid black band, it is a female.

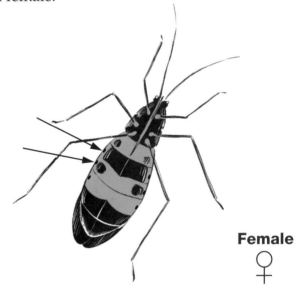

Female
♀

Males tend to be smaller than females. When you see a mating pair, observe closely to see if the female is noticeably larger than the male.

4

A few days after mating, the female starts laying clusters of 20 or more yellow eggs. The clusters are called **clutches.** A female might lay five or more clutches. In nature the female lays her eggs in a ball of milkweed seed fluff or under a bit of bark for protection. In your classroom habitat she will usually lay them in the polyester wool.

After mating and laying eggs, adult milkweed bugs might live another 2 months in a kind of buggy retirement. The life span of the milkweed bug in a sheltered habitat bag in a classroom is about 4 months. In the wild they probably don't live that long.

In captivity a milkweed-bug population will continue to reproduce one generation after another. If the 2-month life cycle continues, six generations could be produced in a year! That's potentially a lot of milkweed bugs.

In the wild, however, milkweed bugs stop reproducing in the fall when the weather gets cold and the milkweed plants die. The adult bugs that have not reproduced find protected places where they can hibernate. Even though they have natural antifreeze in their bodies, winter surely takes its toll on the population. But life is durable, and at least a few bugs always live until spring. And when a male and a female survivor meet on a fine spring day, they naturally continue the process of building up the population during the warm days of summer.

Oncopeltus fasciatus

For purposes of classification, scientists have divided the animal **kingdom** into major groups called phyla. Each **phylum** is divided into classes. Each **class** is divided into orders. Each **order** is divided into families. Each **family** is divided into genera, and each **genus** into **species.** A species is a basic category or a kind of animal. A species consists of individuals that are similar in structure and that can breed to produce offspring. Individuals in a species will not generally breed with individuals from another species.

The milkweed bug studied in this course is a member of the phylum **Arthropoda,** the class **Insecta,** the order **Hemiptera** (true bugs), the family **Lygaeidae** (seed bugs), the genus **Oncopeltus,** and the species *fasciatus.* The common name is large milkweed bug and the scientific name is *Oncopeltus fasciatus.* A close relative of the large milkweed bug is *Oncopeltus sexmaculatus,* the six-spotted milkweed bug (same genus, different species). Another more distant relative of these two bugs is *Lygaeus kalmii,* the small milkweed bug (same family but different genus and species). Each is considered a different species.

LIFE IN A COMMUNITY

You live in a community. But it might not be the community you think you live in. Most people grow up with the idea that their community is the homes, businesses, roads, institutions (schools, hospitals, government offices, police, library, etc.), and people they interact with regularly. Community is people meeting, working, planning, and living together.

That's one meaning of community. But that's not what community means to the ecologist. From a scientific point of view the **community** you live in comprises the populations of plants, animals (including humans), and other living **organisms** that live and interact in an area. Furthermore, community is *only* the populations living in an area, not the place where they live. Your ecological community might include one or more populations of rodents, several populations of trees, lots of populations of grasses, hundreds of populations of insects, and countless populations of microscopic organisms like algae, fungi, and bacteria. Now *that's* your community.

Because communities are described by the organisms living in them, no two are exactly the same. The community of organisms living on an island in Lake Superior is similar to, but not the same as, the community living on the shore of Lake Superior near Marquette, Michigan. Both of those communities, however, are very different from the community of organisms living on the coral reef just offshore from Key West, Florida.

ISLAND	LAKESHORE	CORAL REEF
sparrow	sparrow	algae
crow	crow	sea grass
osprey	osprey	sponge
mosquito	mosquito	coral
black fly	black fly	lobster
badger	deer	moray eel
skunk	badger	angelfish
squirrel	fox	gobi
mouse	skunk	nudibranch
pine	mouse	snail
spruce	moose	sea star
oak	pine	octopus
beech	spruce	sea urchin
blueberry	oak	shark
moss	beech	porpoise
grass	blueberry	shrimp
	grass	

Another factor that defines a community is the interaction among organisms. In a community, every organism's life is connected to every other organism's life in some way.

Some interactions are obvious. When a robin eats a worm, the robin is fed and the population of worms decreases by one. Other interactions are not as easily observed. The importance of an ash tree as a safe place for a robin to build a nest to raise young might be overlooked, but without the protection provided by the tree, the chances of the robin's increasing the population are reduced.

When the robin dies and its body falls to the forest floor, populations of decomposers, like bacteria and fungi, consume the remains, increasing their population and recycling the minerals from the robin's body back into the environment. The ash tree benefits from the mineral nutrients returned to the soil, increasing its vitality. A more vital ash tree is more likely to grow larger and produce seeds to reproduce new ash trees. The robin's mineral remains, processed by the decomposers, nourish the tree, which in turn provides more nesting sites for the next generations of robins.

This is just one peek into the complexity of a community. The interactions among the organisms together with the nonliving surroundings is called an **ecosystem.**

What organisms do you interact with in your community? Which ones do you eat, and which ones eat you? Which ones compete for your food, and which ones provide shelter or comfort?

BIOSPHERE 2: An Experiment in Isolation

Can human beings live in an airtight building for 2 years with no life support except for the organisms they bring inside with them? Can they set up an ecosystem that provides for their every need and the needs of the other populations around them?

These were the big questions that Biosphere 2 was built to answer. In 1984 construction began on the world's largest isolated environment. The location was the Sonoran Desert, a short distance north of Tucson in Arizona. This part of the country experiences a high percentage of days without cloud cover. Sunshine is essential for an experimental ecosystem because it is the only energy source to sustain the humans and hundreds of other species in the closed system.

One critically important factor was sealing the 3.1-acre live-in terrarium from the outside environment. That means no gases or organisms entering or leaving the system. First a 500-ton stainless-steel liner was laid down to isolate Biosphere 2 from the earth. Then a massive glass, steel, and concrete greenhouse was constructed on the steel base to isolate Biosphere 2 from the atmosphere. The $150 million chamber was ready for business in 1991.

Four men and four women were identified as the first team. But before they closed the door, they had to populate Biosphere 2 with other organisms in order to turn the newly finished building into an ecosystem. The ecologists working on the project spent a long time selecting organisms. They knew that they needed plants, animals, and microorganisms. They needed to have food for eight people and life support for the organisms that would provide the food. They needed organisms to refresh the air and dispose of waste materials. The planning was complex and detailed—lives depended on getting it right.

The First Mission

In September 1991 the door was closed and sealed with the eight Biospherians and 1800 other populations on the inside. The challenge faced by the humans was the same one astronauts will probably face during our first visits to other planets. The trip to the Moon takes a few days. A trip to Mars might take a year. The most efficient way to make such a journey would be in a miniecosystem, where everything needed for life recycles.

Biosphere 2 had surprises for the scientists inside. Before long they noted that the oxygen concentration began to drop. The oxygen started at 21%, the concentration of oxygen in Earth's atmosphere, but got down to 14%. This was a dangerous level for the people. Where was the oxygen going?

Analysis revealed that the soil in Biosphere 2 was too rich in organic matter. The populations of microbes were growing out of control, using too much of the oxygen. The scientists reasoned that if the oxygen concentration was going down, the carbon dioxide (CO_2) concentration should be going up. But the concentration of CO_2 was not going up as fast as the scientists calculated. It was later discovered that the CO_2 was being taken up by the massive

amount of concrete that was still curing.

On the biotic side, a problem came up with ants. An uninvited species, known as crazy ants, got into Biosphere 2 somehow and caused disruptions in the community. Not only did the ants put pressure on other organisms in the ecosystem, they clogged vents and chewed on wiring, creating quite a nuisance.

How could tiny organisms like ants cause a major problem in the Biosphere 2? Crazy ants form "super colonies." Super colonies have many queens and many nests. All of the ants work together to search for food, share food, and distribute resources. Most other species of ants form colonies with a single queen in a single nest and are highly territorial toward other colonies of the same species of ant. Crazy ant colonies, on the other hand, cooperate with one another. This gave them an advantage over other species of ants in Biosphere 2. While crazy ants are not aggressive to others of their species, they are very aggressive in searching out and attacking prey. They can effectively communicate the exact location of the prey to other ants in the super colony. Then they can launch an attack that will overwhelm even a large insect such as a cockroach.

Crazy ants, like other ants and many other animals, communicate with each other by using pheromones. Pheromones are scent chemicals that send signals to other animals of the same kind. For example, ants leave pheromones on the ground to mark a trail for other ants in the colony to follow. While most other ants are thought to have only one trail pheromone, crazy ants have at least three different ones. Some of these pheromones evaporate faster than others, so they stay on the trail for only 2–3 minutes, while other pheromones may last for 24 hours. Crazy ants, with more than one pheromone, can provide more information to other ants so the colonies can adjust quickly to changing conditions. The superior communications and the super colony cooperation seem to be the characteristics that gave the crazy ants the advantage over the other insect species in Biosphere 2 and allowed the crazy ants to displace most of the other arthropods.

The Biospherians made it through the 2-year mission, but just barely, and not without a little assistance. The oxygen problem could be solved only by pumping in extra from outside. A second mission in 1994 lasted 6 months. Following the second mission, a decision was made. Biosphere 2 would no longer be a live-in facility, but would be transformed into a unique ecological research center. In 1996 Columbia University took over management of Biosphere 2 and continues to oversee the important international research going on there.

The Current Research

It is fairly easy to do an experiment with the plants and animals in your terrarium or aquarium. You could add some CO_2 to one terrarium but not the other, to see if it made plants grow faster or slower. You could increase the temperature of an aquarium and monitor the health of the fish. But would your experimental results tell you what would happen in the rain forest or the ocean?

How do you study an entire ecosystem? A real ecosystem is much larger and more complex than the ones you can build in class. To try to answer some of these questions, you could add CO_2 gas to a field or forest to see how plant growth might change, but the wind would soon blow the CO_2 away. You could study the weather over a coral reef for a period of 40 or 50 years to see if there are any patterns, but it would be impossible to control variables. You might notice a difference in plant growth and health when you compare a

wet, cold, rainy year to a hot, dry, sunny year. Is the difference because of the rain, the moisture in the air, the temperature, or the amount of sunlight? It is impossible to say with so many variables changing at once.

What you really need is a box like your terrarium or aquarium that is big enough to hold a whole ecosystem. Then you could do experiments and control all the variables except the one you are studying. Where can you find such a box? Biosphere 2.

Biosphere 2 covers almost as much area as four football fields. Inside are seven different environments: rain forest, desert, tropical ocean, marsh, savanna, thorn scrub, and agroforest.

One area of intensive study is the tropical rain forest ecosystem. Tropical rain forests can soak up CO_2, a greenhouse gas that contributes to global warming. Tropical rain forests are sometimes called the lungs of the planet because they take in so much CO_2 and produce so much oxygen during photosynthesis. Dense vegetation and long days of direct sunlight year-round enable rain forest plants to carry out photosynthesis at a higher rate than anywhere else on Earth. Some scientists think rain forests have great potential for controlling the rising CO_2 level.

Another possibility is that more CO_2 in the atmosphere will result in rising global temperatures all over Earth. Warmer temperatures could result in less rain, because less water vapor will cool enough to condense into raindrops. Water and CO_2 are both needed for photosynthesis to take place. If there is less rain, photosynthesis slows down, which means less CO_2 will be removed from the atmosphere. Increased CO_2 heats the atmosphere even more, reducing rainfall even further. Over time this can cause the average global temperature to rise, which can have a significant effect on many ecosystems.

John Adams, senior research specialist, ascends a canopy access system inside the Biosphere 2 Tropical Rain Forest to check leaves sealed in a branch bag.

Inside the rain forest environment in Biosphere 2 scientists are doing experiments to determine how drier conditions affect the amount of CO_2 taken up by rain forest plants. They can control the amount of "rainfall" with the overhead sprinkler system. CO_2 can be pumped in. The temperature can be regulated with huge heating and cooling units. Scientists can change any of these variables one at a time to see what effect each has on the amount of CO_2 taken up by the plants.

What will happen if the CO_2 level and the global temperature continue to rise? Will the rain forests take up more CO_2? Will the rain forests take up less CO_2? How will warmer temperatures or more CO_2 affect the health of the plants? Scientists at

Biosphere 2 hope to answer these and many other questions about how the rain forest ecosystem responds to change.

Coral reefs have been called the rain forests of the sea because there is so much diversity of life in coral reef ecosystems. Coral defines the ecosystem. Algae, fish, crabs, and many other kinds of sea organisms live on, in, and around the coral structures.

Microscopic **phytoplankton** (single-celled algae and photosynthetic bacteria) are essential to the health of ocean ecosystems. Phytoplankton, which conduct photosynthesis, are food for coral and other small animals, which in turn are food for larger sea animals. Phytoplankton are the base of marine food chains just like green plants are the base of terrestrial food chains.

Marine biologists report that there are fewer and fewer fish and other organisms living in the coral reef ecosystems around the world and that the growth rate of coral has slowed. During the winter of 1997–1998, one-tenth of the world's coral reefs died. The temperature of the water in the affected areas that winter was 2–3°C above normal, making it one of the warmest periods on record. But was it temperature alone that killed the coral, or was it more complicated than that?

Several scientists are studying the coral reefs in the ocean biome at Biosphere 2. The Biosphere 2 "ocean" holds 2,500,000 liters of water. The depth ranges from 0 meters (m) at the beach to 7 m in the deepest part. Scientists have investigated several factors they think might affect the health and survival of the coral.

When they varied the concentration of CO_2 dissolved in the water, the health of the corals declined. They found that excess CO_2 dissolved in the water prevented coral from getting calcium out of the water to build their skeleton.

The studies of the coral reefs in Biosphere 2 provide evidence that the changes taking place in the ocean environment such as more dissolved CO_2 affect the coral reef ecosystem in negative ways.

As scientists learn more about ecosystems, two things become very clear. The first is that any change in one part of an ecosystem affects every other part of the ecosystem, many times in ways that no one could have anticipated. The second is that the more we learn, the more we realize how complex natural ecosystems are and how little we understand about the way they work or what effect human activity has on them.

Why should we care? What difference does it make if rain forests or coral reef organisms are disappearing? That's not where we live.

Well, it *is* where we live. Our planet is small. The atmosphere surrounds the planet and the seas wash up on all the continents. Changes in one ecosystem are communicated to the rest of the world by flowing air and water. Everything is connected. Small changes in global temperature can have a huge effect on weather patterns. And weather distributes water, and water is life.

Maybe you can join the small community of people trying to answer some of these tough ecosystem questions. College students from several universities across the country attend classes at Biosphere 2 for a semester to study environmental problems. There is also a summer program for high school students who are interested in studying environmental issues. Maybe in a few years...

More information about these programs is available on the Biosphere 2 website at www.bio2.edu.

The Planetary Spheres

One way scientists think about Earth is as a set of nested, interacting spheres. The **lithosphere** is the rocky, mineral part of the planet that extends from the solid surface into the mantle. This is the hard part of the planet that provides a sense of solidity and stability... most of the time. Periodically, we get reminders in the form of earthquakes and volcanic eruptions that the solid Earth is actually restless and dynamic.

Wrapped around Earth is the **atmosphere,** the thin layer of gases that extends, for all practical purposes, no more than 600 kilometers above the surface. The atmosphere is a source of essential

Scientists gather at Biosphere 2 to conduct rain experiments at the laboratory's ocean biome.

Six inches of fresh water fell into the saltwater ocean over a 2-hour period to measure the effect a freshwater addition would have on the exchange of carbon dioxide into and out of the ocean. The rain and air-water gas exchange experiment was funded, in part, by a grant from the David and Lucile Packard Foundation.

gaseous chemicals, an energy-transfer system, a shield protecting us from extraterrestrial radiation, and an insulator. It is also an important medium for water distribution.

Earth is a water planet. Because of the temperature on Earth, water exists naturally in three states: liquid, gas, and solid. All the water on Earth makes up the **hydrosphere.** The hydrosphere includes the oceans, lakes, rivers, streams, and aquifers. It includes the polar icecaps, glaciers, snowpacks, and permafrost. It also includes the aerial water vapor and condensates in the form of clouds, fog, and precipitation.

And finally, creeping, hiding, running, burrowing, flying, slithering, and swimming through, over, under, onto, and into the other three spheres is the **biosphere.** The sum total of all the living organisms on Earth is the biosphere. It is this raggle-taggle, at times improbable, assemblage of millions of different kinds of life-forms that gives Earth its particular flavor.

All four spheres can be bundled into one global sphere called the **ecosphere.** The ecosphere is that portion of a planet that is inhabited by life. Thus, it includes a portion of the atmosphere, a portion of the lithosphere, a portion of the hydrosphere, and all of the biosphere. We focus on the biosphere in this course. However, we will continually consider the interactions between living organisms and the other three spheres to reinforce the idea that life is never disconnected from the physical environment.

THINK QUESTIONS

1. Give at least two examples of how a change in one variable in an ecosystem can start a chain reaction that affects several other variables.

2. Why is global warming considered by some scientists to be such an important problem?

3. What are some advantages of doing research on ecosystems in Biosphere 2 rather than in the natural ecosystem? What are some disadvantages?

4. Think about the statement "Every decision has an environmental impact." What decisions do you make that add carbon dioxide to the environment? What decisions do you make that would add less carbon dioxide to the environment than you currently add?

5. Why should we be concerned about species becoming extinct? What endangered species are found in the area where you live? What has caused them to become endangered? What is being done to help them survive?

Where Does Food Come From?

I have a sweet tooth for caramel corn...it's one of my favorite snacks. I have a friend who goes for chocolate-covered raisins. Those are both popular snack foods. We eat snack foods when we are a little hungry in the afternoon or maybe at a movie or sporting event.

Snacking is one way we get food, but serious feeding usually takes place at meals—breakfast, lunch, and supper. That might turn out to be cereal, fruit, or eggs in the morning and a sandwich, bowl of soup, or maybe a slice of pizza for lunch. For supper it might be a salad, plate of pasta, some enchiladas, chicken and potatoes, or a bowl of stew with bread. With meals and snacks, it seems like we are eating something just about all the time. Why is that?

We eat to live. Everybody has to eat. For that matter, every animal has to eat to live. Food is the source of two essential elements of life—things that we need to stay alive and healthy. They are energy and chemical building blocks.

Building blocks are matter. That is, they are made out of atoms and are solid "stuff." Building blocks are carbon-based chemicals used to make body tissues, heal injuries, and replace worn parts. Building blocks come from the fats, proteins, and carbohydrates in

food. Without building blocks you couldn't grow or mature as you get older, and you couldn't replace hair or discarded skin cells.

Energy is not matter; it is not made of atoms. Energy is the drive that gets things done. Some familiar forms of energy are heat, light, sound, and electricity. Energy is needed to do all the things that living organisms do. Energy is needed to move, talk, digest food, keep warm, think, grow, and feel. Energy is the drive that keeps the machinery of life running.

How Do Chemicals and Energy Get into Food?

Food is the source of matter and energy for all organisms. But not all organisms eat food. How can some organisms get the benefits of food without eating it? The answer is simple. Some organisms make food right in their own cells. Organisms that make their own food are called **autotrophs** (*auto* = self; *troph* = feed). Autotrophs are self-feeders.

Autotrophs make food from water, carbon dioxide, and light. The process is called **photosynthesis** (*photo* = light; *synthesis* = putting things together). The autotrophs of this planet are the plants, some protists (including algae), and some bacteria. These photosynthetic organisms never need to eat other organisms to survive. They just make food by photosynthesis, bonding carbon atoms together to make sugar molecules. Below is the equation that shows how six water molecules (H_2O) and six carbon dioxide molecules (CO_2) are changed into a glucose molecule (sugar) with energy from light. Note that six molecules of oxygen (O_2) are also released.

$$6\ CO_2 + 6\ H_2O + Light \rightarrow C_6H_{12}O_6 + 6\ O_2$$

The sugar molecules (food) are rich in energy from sunshine and loaded with atoms to use as building blocks. The photosynthetic organisms use the food they manufacture to get the energy and building blocks they need to live. Plants never eat ham sandwiches, algae never eat enchiladas, and photosynthetic bacteria never eat fruit. They don't have to.

How the Other Half Lives

Humans are not autotrophs. We can't make our own food. The same is true for all the other animals, the fungi, most protists, and most bacteria. But all organisms need food, so the ones that don't make their own food eat the ones that do. Organisms that eat other organisms are called **heterotrophs** (*hetero* = other; *troph* = feed).

Humans get their food in thousands of different forms. The great majority of people in the United States get food at grocery stores and restaurants. The caramel corn, pizza, and chicken mentioned earlier are familiar to just about everyone, but where did they come from before they arrived at the store? And what autotroph made the food in the first place? It's fun to find out. And when you trace any food back to its source, it all starts with plants or plantlike organisms. Always.

Let's follow the paths that led to a pepperoni pizza (do you like that with anchovies?). The main ingredients are bread dough, tomato sauce, cheese, and pepperoni sausage. Start with the dough...that's a short path. The pizza dough is bread, made from wheat flour. Wheat flour is made by milling (grinding) the seeds of the wheat plant, a kind of grass. So the crust is a part of the pizza that comes fairly directly from a plant source.

Tomato sauce is made by grinding tomatoes to a pulp and cooking them with a few seasonings. The sauce is a part of the pizza that comes fairly directly from the fruit of the tomato plant.

Cheese is processed from milk. Milk comes from cows. Cows need to eat to live, just like we do. Cows don't eat pizza and chocolate-covered raisins; they eat grass and other plants. The milk is produced by the cows that eat plants to survive. The nutrients in milk come from the food value of the grass eaten by the cows. The cheese comes indirectly from the grass eaten by the cows. If there is no grass, there are no cows, and therefore no cheese for our pizza.

Pepperoni is made from the ground-up muscle of cattle and pigs. In order to have the spicy Italian sausage, there had to be livestock to butcher. The livestock grew to market size by eating plants, probably grasses and grain seeds like corn and millet. The plants nourish the animals that provide the meat that is made into pepperoni for the pizza.

Did you order anchovies? The case of these little fish is different, but only slightly. Anchovies don't eat plants, as do cows and pigs. Anchovies are animals that eat other animals. The animals they eat are tiny free-swimming critters called zooplankton. But what do the zooplankton eat? They eat even smaller plantlike organisms called phytoplankton. What makes them plantlike is that they can make their own food by photosynthesis, just like plants. So the anchovies eat little animals that eat tiny plants to survive. The anchovies arrive on your pizza because of the food value in the microscopic phytoplankton floating in the sea, making food.

You eat to get the energy in food. Where does your energy come from when you eat that pizza slice? The crust and sauce have a

bit of energy that wheat and tomato plants captured from the Sun. The cheese and sausage come from animals that ate plants that captured the Sun's energy. The anchovy filet comes from an animal that ate an animal that ate the tiny phytoplankton that captured the Sun's energy.

Life runs on energy from the Sun. The Sun's energy is captured in food by photosynthetic organisms. Can you figure out how the Sun's energy got into caramel corn and chocolate-covered raisins? What's *your* favorite food? How did the Sun's energy get packaged into that food?

TROPHIC LEVELS

Energy and Life

All organisms need energy in order to survive. How an organism acquires its energy defines its role in the ecosystem. But first, what is energy?

Energy is the ability to do work. Living organisms need energy to perform the basic functions of life, such as growth, reproduction, gas exchange, elimination of waste, getting water and nutrients, and responding to the environment. These processes take place in each and every individual cell, whether the cell is a free-living organism (like an amoeba) or one of millions within a multicellular organism. All cells require energy because all cells are constantly working.

Organisms use energy, but they don't make it. Energy must be obtained outside of the organism. Energy is usually obtained from the environment or from another organism that got it from the environment. There are millions of different organisms on Earth, but each obtains energy in one of the following ways, regardless of its size or complexity.

Producers

The Sun is the source of energy for just about all the organisms in every ecosystem on Earth. Some organisms, notably plants, algae, and phytoplankton, capture energy from the Sun using **photosynthesis** (*photo* = light; *synthesis* = putting things together).

Photosynthesis is the chemical process that turns carbon dioxide and water into carbohydrates in the presence of sunlight and chlorophyll (a green pigment found in the cells of plants and other photosynthetic organisms). Oxygen gas is a by-product during this process. Because photosynthetic organisms produce carbohydrate molecules, which have energy in the chemical bonds that hold them together, they are called **producers.** Producers create **biomass** (*bio* = living; *mass* = matter). Biomass is the total organic matter in an ecosystem.

Producers convert light energy into chemical energy. The chemical energy is in the carbohydrate molecules that photosynthesis produces. Carbohydrate is one form of food. Cells use the energy in food to do the various kinds of work they need to do. Because they make their own food, producers are also referred to as **autotrophs,** or self-feeders (*auto* = self; *troph* = food).

Producers on land include plants such as grasses, shrubs, and trees. Aquatic and marine producers include microscopic phytoplankton, water plants, sea grasses, and seaweeds such as giant kelp.

Consumers

Producers transform the Sun's energy into food energy. On land the main producers are plants. They use the food energy to stay alive. Not only plants, however, use the food energy they make. Other organisms eat plants to get the energy they need for life. Organisms that eat other organisms are called **consumers.**

Consumers cannot make their own food, so they must obtain energy by eating other organisms. Organisms that eat other organisms are also referred to as **heterotrophs** (*hetero* = different).

Animals that eat only plants, such as mice, elephants, deer, and caterpillars, are called **herbivores.** They are also known as **primary consumers,** because they are the first level of consumer.

When primary consumers are eaten, they provide food (energy) for other organisms, called **secondary consumers**—organisms that eat the consumers that eat the producers. Secondary consumers are **carnivores,** or meat eaters. They live by eating herbivores. Secondary consumers include wolves, spiders, snakes, and hawks.

Tertiary consumers, animals that feed on the secondary consumers, may be part of the ecosystem. Examples of tertiary (or third-level) consumers include sharks, pelicans, polar bears, barracudas, and orcas (also known as killer whales, the world's largest porpoise).

Some animals, such as brown bears, crows, chimpanzees, and most humans, have a diet of both producers and consumers and are called **omnivores.**

Decomposers

Many producers and consumers die without being eaten by other organisms. Their biomass falls to the ground or bottom of the ocean or lake and is available for other organisms to eat. Organisms such as bacteria, mushrooms, and other fungi are called **decomposers,** because they eat the remains of dead organisms.

When something rots or decays, it is actually being consumed by decomposers. This process of decay transfers every last bit of energy from the dead organism to the decomposers. When the decomposers are through, all that remains is a few simple chemicals, which return to the soil or water. These simple chemicals are minerals, which can be used once again by the producers to begin the cycle. Without decomposers, the ecosystem's recycling crew, Earth would be buried under mountains of wastes and dead organisms!

Bacteria and fungi are the most important decomposer organisms, and their action in the ecosystem clearly satisfies the description of decomposer discussed earlier. But what about organisms like worms, maggots, termites, and even vultures and coyotes, which consume dead organic matter for the energy they need? Are they decomposers or are they herbivores and carnivores? Ecologists are still defining the roles played by the less glamorous, but critically important, organisms that clean up the dead and discarded bits of life. A thriving community of organisms works on the film of dead organic matter that covers much of the surface of Earth at the interface between land and air or land and water. This organic matter, known as **detritus,** is the home of the **detritivores,** organisms that eat dead material. Not all detritivores reduce the matter to simple chemicals. Worms and beetle larvae, for instance, miss a significant amount of matter, and produce feces, which have energy that can be used by more thorough decomposers.

In this course we cast the bacteria and fungi in the role of decomposer, because they are the organisms that extract the last iota of energy from the matter they exploit. Macroscopic "decomposers," such as earthworms and other scavengers, are considered detritivores or consumers.

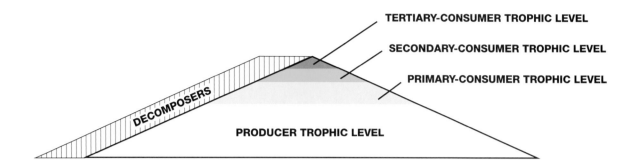

TERTIARY-CONSUMER TROPHIC LEVEL

SECONDARY-CONSUMER TROPHIC LEVEL

PRIMARY-CONSUMER TROPHIC LEVEL

DECOMPOSERS

PRODUCER TROPHIC LEVEL

Organizing the Ecosystem Based on Feeding Relationships

Organisms in an ecosystem can be grouped by how they get food. The groups are called **trophic levels.** Trophic levels are feeding levels. The trophic level at the base of the ecosystem is the producers. The producer trophic level always has the largest amount of biomass. The next largest amount of biomass is in the trophic level made up of primary consumers. The primary consumers are followed by the secondary consumers, and so on.

When the groups are placed in order from producers to primary consumer to secondary consumers, and so on, the result is a layered representation of the ecosystem. It is often represented as a pyramid, because each level has less biomass than the one below it.

It's difficult to place the decomposers in the pyramid model of trophic levels because they affect every trophic level. They are often placed along the side to suggest that they interact with all the other trophic levels.

If you were organizing the organisms in an ecosystem into trophic levels, what would you do with the raccoons and crayfish? They both eat plant material, so they go in the primary-consumer trophic level. But they both eat insects and other animals, too. That would place them in the secondary-consumer or, possibly, the third-level-consumer trophic level. Where do you place them?

Animals that are generalists, that is, those that feed at several trophic levels, should probably be represented in each trophic level where they play a role. Thus, the crayfish might appear in three levels because it eats plants, insects, and fish.

Food Webs

In a typical ecosystem, the producers, consumers, and decomposers transfer food energy from organism to organism to form what is known as a **food web.** A food web is an informative way to represent energy flow through an ecosystem because it shows all the feeding relationships. For instance, grass, lupines, and pine trees are producers. Voles, pocket gophers, and chipmunks are primary consumers. Foxes, coyotes, and barn owls are secondary consumers in an ecosystem.

barn owl	coyote	fox
gopher	vole	chipmunk
lupine	grass	pine

The same organisms can be organized into a food web. A food web uses arrows to connect organisms that eat one another. The arrow points from an organism to the organism that eats it. Another way to think of it is that the arrows point in the direction that the energy in the food goes. For instance, when a spider eats a fly, the food energy in the fly goes into the spider. The arrow points from the fly to the spider.

This is how our little community of organisms can be organized into a food web.

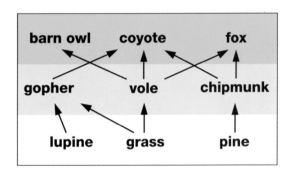

If the organisms are grouped by trophic level before arrows are drawn, a picture of community interactions is easier to see. Another important fact can be seen: some organisms must die to maintain the lives of others.

Energy Transfer

What becomes of the food made or eaten by an organism? Much of the chemical energy available in the food made by a producer or eaten by a consumer is used to maintain life functions. A portion of the energy is stored in the body of organisms in the form of biomass, but most is used to do work and then passes to the environment in the form of heat. So the energy needed to run, think, digest, pump blood, and perform all other life activities passes through the organism and into the environment.

The energy that passes through the ecosystem (from the Sun to the environment) is often referred to as "lost" because it is not stored as biomass to be used as food at the next trophic level. Only a small percentage (an average of about 10%) of the food consumed at a trophic level is converted to biomass. As a result, consumers must eat a lot just to maintain their body mass and functions.

What does the 10% rule tell us about the distribution of organisms in the various trophic levels in a typical ecosystem? Will there be more producers? More primary consumers? More secondary consumers? The 10% rule tells us that the biomass of producers will be much larger than the biomass of primary consumers. Using the 10% rule again, the biomass of primary consumers will be much larger than the biomass of secondary consumers.

Food Pyramids

At each trophic level, the amount of energy transferred to new biomass through growth and reproduction is only about 10% of what the organism eats. In other words, there are usually fewer predators than prey. In the same way, because so much energy is lost as primary consumers eat producers, there are generally fewer herbivores than plants. This is why trophic-level diagrams are often pyramids. Each layer is smaller as you go up the trophic levels, so the ecosystem diagram tapers to a point on top.

It is extremely rare for an ecosystem to have fourth-level consumers that eat third-level consumers because the amount of energy lost in each transfer is so great. Most examples of fourth-level (and higher) consumers live in freshwater aquatic or marine ecosystems. An orca is an example of a very high-level consumer. An orca might eat a sea lion, which eats a salmon, which eats an anchovy, which eats zooplankton, which eats a single-celled alga. The orca is a fifth-level consumer. And when the great white shark eats the orca...you get the idea.

Summary

Living organisms are complex. Their bodies are made of atoms. That is the biomass of life. The functions of living organisms are driven by energy. The energy for life comes from the Sun, captured in energy-rich molecules. The energy-rich molecules are called food.

Every organism needs a constant supply of atoms and energy. Autotrophs (producers) get atoms and energy from raw materials in the environment. Heterotrophs (consumers) get atoms and energy by eating other organisms. Atoms and energy move up through the trophic levels in an ecosystem by feeding relationships.

Dead organic matter still has valuable atoms and energy. Decomposers get the last bit of energy out of organic matter and reduce the atoms to simple chemicals. The energy that entered the ecosystem as sunlight leaves the ecosystem as heat. The atoms that entered the ecosystem as food made by producers return to the environment to be used by living organisms again.

Energy passes through the ecosystem only once. Matter recycles again and again and again. The simple chemicals from which life is constructed—water, carbon dioxide, minerals—can enter and reenter the life process time and time again. Matter recycles.

Energy, on the other hand, comes from the Sun, passes from organism to organism for various periods of time, and then is gone. Energy passes through once and is lost to the life process. Energy transfers but does not recycle. The fate of virtually all energy that passes through the trophic system is to be radiated into the environment as heat. Once released to the environment, it is gone.

Limiting Factors

Nothing lives forever. Even the ancient bristlecone pine trees of the high mountains of the western United States die after a few thousand years. Most organisms live much shorter lives. Many insects live a few months; fish and small mammals a few years; many plants, reptiles, birds, and large mammals a few decades; and a scattering of others, like trees, a few centuries. Life is a temporary thing...for the individual.

If a species is to continue to exist on Earth, the species must produce new individuals continually. Producing new individuals to maintain a population is **reproduction,** and every species has a way of reproducing.

The rate at which a species can increase its population is its **reproductive potential.** Some species, like elephants, have modest reproductive potential. A female elephant reproduces a single offspring every 4 years. A single female Atlantic cod, on the other hand, can lay 10 million eggs a year. Clearly the potential for the cod to increase its population is much greater than the potential for the elephant to increase its population.

So why don't populations spiral out of control? Why aren't there billions of billions of trillions of Atlantic cod filling all the oceans from top to bottom after 5 or 10 years? Because there are **limiting factors** imposed on every population on Earth. Limiting factors control the sizes of populations.

BIOTIC LIMITS: PREDATION

One way that populations are limited is through **predation.** Every organism is desirable to some other organism as a source of food. As we know, food provides the energy that is essential for survival. Therefore, if a species reproduces a lot of biomass, it will attract predators to take advantage of the energy source. We see this kind of population control in Mono Lake when the brine shrimp feed on the planktonic algae, reducing their numbers, and in turn the brine shrimp are eaten by phalaropes and gulls, reducing the population of shrimp. Predation can occur at any stage in the life cycle of an organism, including eggs and seeds, young, mature, and old. Populations are limited by removal of individuals as they are eaten.

Diseases limit populations in the same way. Even though we don't usually think of a large animal or plant being attacked by a microscopic bacterium, the result can be the same. A mountain lion capturing a deer, or a hawk taking a squirrel, removes an individual organism from the population. A disease organism can enter a population and kill many organisms, which also limits the size of the population.

BIOTIC LIMITS: RESOURCES

Populations are limited by food supply. If an organism cannot acquire the energy needed to survive and reproduce, the population will decline and, with it, the potential for producing the next generation. If a snake cannot find enough mice to sustain itself, it will starve to death. Even if it survives, it may be so weak that it can't reproduce. Similarly, if there is a poor crop of acorns, squirrels may starve. Even if they survive, they may not be able to feed their young. In 1982 a reduced population of brine shrimp in Mono Lake prevented the California gulls from successfully feeding their chicks. Most of the gull offspring died that year. Lack of food is one of the most important limitations on populations.

ABIOTIC LIMITS: REPRODUCTIVE ENVIRONMENTS

Many organisms require specific conditions in order to reproduce. If the number of locations where reproduction can occur or their quality is limited, reproduction will be limited. Bank swallows need sandy cliffs in which to dig nesting burrows. If a sandy cliff tumbles down during a flood or earthquake, suitable nesting sites are lost. Salmon need clean gravel streambeds in which to lay their eggs, and black bears need winter dens in which to give birth. Without an environment that provides for the physical conditions needed to reproduce, young will not be born. Lack of access to required reproductive environments for a species limits populations.

ABIOTIC LIMITS: SEASONS

Seasonal changes put pressure on populations. In the temperate and polar latitudes, winter is a major factor in population limitation. During winter, days are shorter, so primary production by photosynthetic organisms slows or, in the case of deciduous trees, stops entirely. Often winter brings rain, snow, and wind, each of which adds stress to populations.

Some animals respond to the threat of wind, flood, and freezing by leaving the area. Birds, because of their mobility, are famous for migrating to warm regions. Others, like the American bison and caribou, go on long treks to find greener winter environments. Some organisms become dormant, basically shutting down until spring. Frogs, fish, bears, squirrels, snakes, maple trees, and hosts of other organisms use dormancy, reduced activity, and winter sleep to wait out the winter.

These strategies work if a number of conditions have been met.

- The wintering place offers sufficient protection.

- The organism has accumulated enough fat or has stored enough food to survive the winter.

Winter is the main limiting factor for many temperate and polar populations. Many populations decline to minimal levels, like the brine shrimp in Mono Lake, and then expand rapidly in the spring. Seasonal fluctuations in population size such as those at Mono Lake are normal and healthy.

CARRYING CAPACITY

When you stand back and take the large view of life on Earth, you realize it is a struggle to survive there. Every living thing has fundamental requirements for life, and if it doesn't get those things, it dies. One of the most critical requirements is energy.

Energy enters the ecosystem as sunlight. Photosynthetic organisms capture the energy and transform it into carbohydrates, like sugar, that we call food. The energy is in the chemical bonds. The amount of food that can be produced is limited by several factors, including access to light, space for living, and availability of resources such as water, carbon dioxide, and minerals. For any given ecosystem there is a limit to the amount of food that the producers can make.

We know that the other populations in an ecosystem acquire energy by eating each other. Primary consumers eat producers, secondary consumers eat primary consumers, and so on. The number of consumers is limited by the amount of production.

The total number of individuals of a population that can be sustained indefinitely by an ecosystem is the **carrying capacity** for that species. For instance, a backyard ecosystem might support three rabbits year after year on the amount of grass and other vegetation growing there. The carrying capacity for rabbits is three. If six rabbits move in, the carrying capacity of the ecosystem is exceeded. As a consequence, in order to survive, the rabbits will eat so much of the vegetation that they will damage the ability of the producers to produce in the future. Exceeding the carrying capacity of an ecosystem always produces changes that will alter the nature of the ecosystem.

In an ecosystem the consumers never eat all the organisms they prey upon. Squirrels never eat all the acorns, caterpillars never eat all the oak leaves, mountain sheep never eat all the grass, sharks never eat all the seals, and so on. This is very important because if they did, the prey species would be gone. The predators' offspring would have nothing to eat, and the predator

population would die off. In healthy ecosystems there are always survivors of every kind in sufficient numbers to reproduce and keep the population going.

Usually the primary producers establish the overall carrying capacity of an ecosystem. How much food energy is produced by the photosynthesizers that can be consumed and distributed throughout the food web of the ecosystem? When you know the answer to that question, you are closer to knowing the carrying capacity of the ecosystem.

Mono Lake is an ecosystem with a tremendous carrying capacity for its size. Mono Lake has plenty of light, water, carbon dioxide, and minerals. The algae reproduce rapidly. The limiting factor for Mono Lake algae is one element—nitrogen. Even so, the biomass of algae produced in the lake supports trillions of brine shrimp and brine flies. These in turn nourish millions of birds and a few coyotes. A lot of life flows through Mono Lake each year.

Contrast this with a grassland on the Great Plains. Grasses grow more slowly, thus taking longer to regenerate their biomass. The amount of grazing by insects, rodents, deer, and cattle must not exceed the capacity of the grasses to regenerate. The carrying capacity of the grassland is less than the carrying capacity of Mono Lake.

Mono Lake in the Spotlight

Mono Lake is desolate. Huge expanses of desert stretch out from the lake in three directions, and the mighty Sierra Nevada rise up on the fourth side. It's freezing cold in the winter and dry and windy in the summer. Sagebrush and desert grasses come right down to the salt-crusted shore. Mono Lake is salty—far saltier, in fact, than the ocean.

Mono Lake is a unique and ancient ecosystem. It is at least 760,000 years old, making it the oldest lake in the United States. The lake is a tiny leftover puddle, really, representing a huge complex of lakes that covered a bit of California and a large part of northern Nevada and Utah hundreds of thousands of years ago. When the huge lake system dried up, only Great Salt Lake, Mono Lake, and a handful of other small basins retained water. The minerals dissolved in the lake water concentrated in the small lakes, resulting in high salt concentrations.

Over the thousands of years that the lake system existed, it was undoubtedly a place where huge numbers of migrating birds stopped to feed and refresh themselves before continuing their journey. As time passed, the lake receded and the land became arid. The migrating birds came to depend on the small pockets of water more and more. Today the last remaining locations that hold water are essential to the survival of the birds that cross the deserts surrounding Mono Lake.

Even though it looks barren and deserted at first glance, Mono Lake is actually one of the most productive ecosystems on the continent. Below the water's surface a huge population of algae is the base of a food web. Massive populations of brine shrimp and brine flies feed on the algae. These crustaceans and insects in turn provide nourishment for populations of birds that come to the Mono Lake Basin from as far away as the Arctic Circle and the equator. Millions of birds visit Mono Lake between midsummer and fall, feasting on the abundant food. California gulls and snowy plovers nest on the two protected islands.

When you get down to the lakefront, gray bumpy structures, called tufa towers, dominate the Mono Lake landscape. These tufa towers form when calcium-rich water from springs beneath the lake bubbles up through the salty water. Chemical reactions occur between the calcium and salt, slowly building the tufa towers.

Tufa towers grow only underwater. So why are they sticking up all around the shore of Mono Lake? The answer to this question is found in the recent history of changes in this important and unique ecosystem.

Location of Mono Lake

Lakeside view of tufa towers

Mono Lake, the largest lake completely in California (183 square kilometers) (70.65 square miles), sits in a broad basin with no outlet. Water flows into the lake from

several small streams, but no water ever flows out. Consequently, over the years salt has continued to build up in the lake. Mono Lake Basin is in an active volcanic area of California. The two islands, Negit and Paoha, are actually small volcanic cinder cones sticking out of the lake. Paoha, the larger island, still spouts steam from vents, and there are several hot springs on the island.

The most important abiotic factor in Mono Lake is the high salt content of the lake water. In fact, Mono Lake contains about 280 million tons of dissolved salts. This makes the water alkaline. Alkaline water is basic (as opposed to acidic). The alkali salts raise the pH to about 10, and the water feels slippery between your fingers. (For comparison, pure water is neutral and measures pH 7, and household vinegar is acidic and measures pH 3.)

Rain and snow are critically important to the stability of Mono Lake. Average annual precipitation is around 35 centimeters (cm). This isn't a lot. Fresh water coming into the lake is essential to offset the water lost to evaporation. If less water comes into the lake than that lost to evaporation, the lake will shrink, but the amount of salt will not. As a result, the lake water will get saltier.

This is exactly what happened between 1941 and 1981. Water was diverted out of streams that delivered fresh water to Mono Lake, and the lake level declined steadily for 40 years. By 1981 the volume of the lake was reduced by half and the salt concentration doubled. The water level in the lake dropped 13.7 meters (m) (45 feet), exposing hundreds of tufa towers that had been hidden below the lake's surface.

Life in Mono Lake

Although Mono Lake is home to none of the organisms you might expect to find in a lake, like fish and frogs, it is filled with life. The base of the food web is the microscopic planktonic algae. During winter and early spring, algae reproduce impressively. By March the lake is "as green as pea soup" from densely concentrated single-celled photosynthetic algae.

This is when countless brine shrimp hatch out of their hard-shelled cysts on the

Brine shrimp

bottom of the lake. Microscopic at first, the baby shrimp feed eagerly on the planktonic algae, and after a few weeks they mature and produce a huge second generation of brine shrimp. In the months of June and July, 4 or 5 trillion mature brine shrimp, each about 1 cm long, fill the lake.

At the same time, the larvae of the brine flies (also known as alkali flies), which have been dormant on the bottom of the lake, become active and start eating the

Brine flies

benthic algae that grow on the bottom of the lake. By midsummer, countless millions of the brine flies are skittering across the surface of the lake as well.

The bounty of shrimp and flies is eaten by huge numbers of birds. Nesting birds consist of 50,000 California gulls (85% of California's breeding population and the second largest colony in the world) and 400 snowy plovers (11% of the state's breeding population). Migratory birds include 1–2 million eared grebes (30% of the North American population), 80,000 Wilson's phalaropes (10% of the world population), 60,000 red-necked phalaropes (3% of the world population), and smaller numbers of 79 other species of waterbirds.

Between April and October there is constant bird activity. With the departure of the last migrating birds at the end of October, Mono Lake is, for a brief time, quiet. The surface is devoid of life—no flies, no birds. The lake water is clear and still. The brine shrimp ate almost all the algae, and in turn were eaten by the migratory birds. The brine shrimp egg cysts and the dormant larvae of the brine flies lie waiting on the lake bottom. The chill of autumn closes in, and Mono Lake gets cold. Then the cold surface water sinks to the bottom, causing nutrient-rich water from the bottom of the lake to rise to the surface for the 760-thousandth time, and the few remaining algae start to reproduce. The cycle repeats.

Humans in the Mono Lake Ecosystem

There is really only one issue at Mono Lake...fresh water. Life depends on it in Mono Lake as surely as it does in every other ecosystem on Earth. Humans have a tremendous thirst for water, and whenever they see an opportunity to gather some up and apply it to human use and enterprise, they usually do.

The Los Angeles Department of Water and Power (LADWP) looked at the streams running out of the Sierra Nevada into what seemed to be a useless, salty lake and contracted to build a system of diversion dams to ship the water south to Los Angeles. The project was completed in 1941, and virtually all the water destined for Mono Lake was piped to southern California. Without the annual inflow of fresh water, the lake began to dry up. The surface of the lake was at 1956.5 m in 1941. Twenty

years later the lake surface stood at 1950 m, a drop of 6.5 m since the diversion.

By the mid-70s the lake level was down more than 12 m, and the lake held half the amount of water it had before diversions started. Effects started to show in the ecosystem. The salt was twice as concentrated, and this stressed the populations of primary consumers, the flies and brine shrimp. The shrimp particularly were not growing as large as usual, and their numbers were declining. In 1982 the brine shrimp production was so reduced that the 50,000 breeding California gulls were not able to catch enough shrimp to feed their offspring, and 25,000 half-grown chicks starved to death. A generation of California gulls was lost. Furthermore, the water was so low that a land bridge developed, connecting Negit Island to the mainland. The bridge allowed predators like coyotes to walk to the nesting area, where they ate the gull eggs and chicks and drove the adults away.

Mono Lake with land bridge

28

In 1978 a young man named David Gaines became concerned about the rapidly changing conditions in the Mono Lake ecosystem and founded an action group called the Mono Lake Committee. He started to work on ways to reverse the damage done to the Mono Lake ecosystem by the water diversions. He worked tirelessly with the National Audubon Society, California Trout, the California Department of Fish and Game, the U.S. Forest Service, and the LADWP to find solutions to the problem.

In 1994, after years of negotiations, a landmark California Supreme Court decision ruled that the LADWP should release enough water to restore the lake to a level of 1948 m and enough water after that to stabilize Mono Lake at that level. This is expected to take 10–20 years. While 1948 m is below the prediversion level, it represents a compromise between the needs of natural ecosystems and the needs of people. This level should provide some water for Los Angeles, guarantee the vitality of the Mono Lake ecosystem, prevent a land bridge from reaching Negit Island, allow trout to spawn in the freshwater creeks feeding Mono Lake, and ensure excellent scenic views for Mono Lake visitors.

David Gaines

Mono Lake Visitor Center

Arctic National
Wildlife Refuge

Yellowstone
National Park

Mono Lake

Delaware Water Gap
National Recreation Area

Monterey Bay National
Marine Sanctuary

Cimarron
National
Grassland

Monongahela
National Forest

Saguaro
National Park

Everglades
National Park

Florida Keys National
Marine Sanctuary

El Yunque Caribbean
National Forest

Ecoscenario Introductions

Every place on Earth has defining environmental conditions. Some of the most important environmental factors include sunshine, water quality, wind, temperature, air quality, substrate, and chemicals. Because the United States extends from the arctic to the tropics, there are amazingly different conditions throughout the country. The United States may be the most environmentally diverse country in the world.

Because of its environmental diversity, the United States also has incredibly diverse organisms. Something lives wherever you look—on the land, in the water, on mountain peaks, on desert sands, and everywhere else.

Considered together, a physical environment and the community of organisms living in that environment define an ecosystem. The physical conditions and organisms in one kind of ecosystem are different than the conditions and organisms in another kind of ecosystem. There are dozens of different ecosystems in the United States. We have selected 11 for study in this course. The ecosystems and the issues associated with them are called **ecoscenarios.** The ecoscenarios focus on national parks or sanctuaries. Their locations are indicated on the map on page 30, and brief introductions to ten ecosystems are presented on the next ten pages. (Mono Lake is introduced on pages 25–29.)

Some organisms are flexible in their lifestyles. They can survive in a number of ecosystems. The coyote is one example.

Coyotes have a wide range of tolerance for temperature, can use lots of food sources, are fast runners, and are secretive and social. These characteristics allow them to live in forests, grasslands, and deserts.

Saguaro cacti, on the other hand, are specialized for one set of physical conditions. They require hot, arid conditions, with little rain and a bit of shade during their first decade or so of life. They can survive only in the conditions found in the Sonoran Desert.

One species of organism is unique—*Homo sapiens.* Humans live in all the ecoscenario locations. If the environment is unsuitable for humans, like the extreme cold of the arctic or the depths of the offshore oceanic waters, we take a suitable environment into the ecosystem to sustain ourselves. Warm clothes and housing, imported food and water, and a self-contained air supply make it possible for us to live in the arctic and underwater. When humans interact in new ecosystems, for living space, food production, resource acquisition, or waste disposal, we are intruding on an ecosystem inhabited by other organisms.

Human actions in ecosystems raise issues. The benefits for humans must be measured against the impacts on the health of the ecosystems. Our actions must be guided by decisions that take into account the long-term well-being of all concerned, the existing ecosystem and the needs and desires of people. The decisions can be difficult, but they should not be based purely on short-term gains for humans.

Arctic National Wildlife Refuge

migrate to the coast to bear young and feed on the low plants and lichens.

A long-standing issue in the Arctic National Wildlife Refuge is whether the coastal plain should be developed for oil drilling. Other areas in Alaska, such as Prudhoe Bay, have already experienced oil drilling. Petroleum scientists have examined part of the Arctic National Wildlife Refuge, called Area 1002, and predict that there is oil there. When the U.S. Congress founded the refuge, it also authorized future oil development in the northern part of the refuge. People have been debating the issue of oil and gas drilling in Area 1002 for almost 40 years.

The Arctic National Wildlife Refuge in the northeastern corner of Alaska is one of the most pristine, undisturbed places on Earth. To the south are the rugged mountains of the Brooks Range.

The most productive area of the refuge, and the most used by wildlife, is the 600,000-hectare (1.5-million-acre) coastal plain. This area is dominated by an ecosystem known as middle arctic tundra. Here the treeless landscape is covered with low-growing plants over a layer of permanently frozen soil called permafrost. In summer the region is dotted with standing water. During this short, soggy growing season, insects flourish, supporting millions of migratory waterfowl. Thousands of caribou

Cimarron National Grassland

For thousands of years the section of the country between the Rocky Mountains and the Mississippi River was an almost endless sea of grass. The prairies and plains grasslands supported a diverse community of animals, including insects, birds, rodents, and large grazing animals like antelope and bison. Most of the prehistoric grasslands have been converted to agriculture.

Cimarron National Grassland is 44,500 hectares (110,000 acres) in the southwest corner of Kansas. More than a hundred years ago, these lands were known as the Point of Rocks ranch. The Beaty brothers, who operated this ranch, grazed cattle on the plentiful grasses. Around 1885, homesteaders began to settle in this area as well. Years of cattle grazing and farming degraded the soils. In the 1930s strong winds swept through the area, blowing away the topsoil. Cimarron was in the dust bowl that lost millions of acres of grassland soil.

In 1937 the U.S. government started a program to restore the soils. Most of the restored soil is now productive farmland. One large section, the former Point of Rocks ranch, was set aside as the Cimarron National Grassland in 1960. Cimarron is managed to maintain it as a native grassland to serve as a reminder of what once was a major ecosystem in the United States.

The main management issue in Cimarron National Grassland is rangeland fires. Some people view fire as beneficial to the ecosystem and believe it is a tool for management. Others feel fire is dangerous and should be put out quickly.

Delaware Water Gap National Recreation Area

ecosystem from commercial development. In 1978 part of the Delaware Water Gap National Recreation Area was designated a national wild and scenic river. After years of public discussion, plans for a dam and reservoir on the river were abandoned in 1992.

The Delaware Water Gap National Recreation Area is a 64.3-kilometer (40-mile) stretch of the Delaware River, running between the states of New Jersey and Pennsylvania. The Delaware River is the largest free-flowing river in the eastern United States—no dams block the river's flow to the ocean.

The Delaware Water Gap National Recreation Area includes riparian woods along the Delaware River and covers 28,000 hectares (69,000 acres) of eastern hardwood ecosystem in New Jersey and Pennsylvania. This park was established on September 1, 1965, for public recreation, to preserve scenic and scientific resources, and to protect the

The Delaware River is one of the cleanest rivers in the United States. The

Delaware Water Gap National Recreation Area is an example of the meeting of several ecosystems—a freshwater ecosystem, a riparian ecosystem (the trees and other native vegetation that border the river), and a forest ecosystem.

Water quality in the Delaware River is a constant issue. The health of the ecosystems depends on a continuous supply of clean water. Pollution management is essential. Acid rain entering the watershed from remote sources also affects ecosystem health.

El Yunque Caribbean National Forest

Puerto Rico is an island in the Caribbean Sea southeast of Florida. The Caribbean National Forest, on the eastern end of Puerto Rico, is commonly called El Yunque (el•YOONG•kay). El Yunque gets its name from the cloud-shrouded mountaintops. Those same clouds provide abundant, warm rainfall all year, producing a lush tropical rain forest. El Yunque has abundant, diverse vegetation, which supports populations of unique birds and amphibians.

El Yunque has a long history as a tropical forest reserve. In 1876, when Puerto Rico was still under Spanish rule, King Alfonso XII of Spain proclaimed El Yunque a forest reserve. He did this not to preserve its diversity and beauty, but because the forests were filled with trees that were used to build ships. Because El Yunque is a reserve, the forest was used sparingly, and it was not destroyed for cities and agriculture.

The primary concern for tropical forests, including El Yunque, is how to deal with habitat loss and destruction in the past, present, and future. Worldwide, rain forests cover 2% of Earth's surface, yet contain half of all plant and animal species. The rain forests, and the species that live in them, are being lost very rapidly. It is estimated that each hour about

3600 hectares (9000 acres) of rain forest are cleared. At the same time, six plant or animal species go extinct.

Everglades National Park

the Everglades because of his efforts to preserve this habitat. In 1928 Coe wrote to the director of the National Park Service, proposing that some of the south Everglades become a national park. Luckily, the director of the National Park Service had already been thinking that some of this ecosystem should be designated as a park. Everglades National Park was established in 1947. It is the largest designated wilderness area east of the Rocky Mountains and covers 610,684 hectares (1,509,000 acres). The Everglades area is very flat, with the highest elevation only 2.5 meters (8 feet) above sea level.

Everglades National Park is a subtropical wilderness on the southern tip of Florida. The wet prairies of saw grass, sometimes called the river of grass, give this ecosystem its name. The Everglades is a slow moving, very shallow, very wide river in which saw grass grows. This river starts at Lake Okeechobee and flows south to the Atlantic, Florida Bay, and the Gulf of Mexico. The Big Cypress Swamp borders the Everglades on the west, and to the east there is a low coastal ridge.

Ernest F. Coe is sometimes called the father of

Water use and water quality are the primary issues concerning Everglades National Park. Water diversion for agriculture and development, especially in the last 50 years, has reduced the Everglades ecosystem to less than half its original size. Other issues include mercury pollution, endangered species, and introduced species.

Florida Keys National Marine Sanctuary

The Florida Keys are 1700 islands. They are the high points of a huge coral reef system that begins at the tip of Florida and curves southwest for 202 kilometers (km) (126 miles). It ends 145 km (90 miles) north of Cuba.

Surrounding the keys is Florida Keys National Marine Sanctuary. This marine sanctuary covers 9600 square km (2800 square nautical miles). The reefs in the sanctuary form the third largest system of coral reefs in the world. The warm clear water ranges in depth from 0.6 to 610 meters (m) (2 to 2000 feet) with an average of 15.25 m (50 feet).

The reefs of Florida Keys National Marine Sanctuary are biologically diverse and extremely productive. The coral structure provides substrate for algae and habitat for fish, worms, and other marine animals. The Florida Keys also have beds of turtle grass and mangrove forests. These communities provide important nursery habitat for marine fish and other animals.

One of the world's major shipping routes passes along the Atlantic side of the Florida Keys. The keys attract thousands of visitors who enjoy diving, boating, and fishing on and around the reefs. There is danger of pollution from boats and recreational facilities throughout the keys. The delicate coral organisms, which build the reef, are threatened by the intense use of the reef areas.

Monongahela National Forest

acres) near the Monongahela River. This land became Monongahela National Forest on April 28, 1920.

Today, the Monongahela ecosystem is primarily a second-growth forest of more than 75 species of trees, such as black cherry, oak, hemlock, and poplar. It is a popular vacation area, positioned within a day's drive of one-third of the population in the United States.

Monongahela National Forest is located in the Allegheny Mountains of West Virginia. This national forest covers over 363,500 hectares (909,000 acres), the fourth largest national forest in the northeast United States. The landscape is rugged with spectacular views of exposed rocks, spring wildflowers, and colorful fall leaves.

In the 1880s the Allegheny Mountains were logged extensively. Clear-cut logging removed all vegetation in many areas. Major forest fires added to the amount of deforested land. Soil erosion was widespread in the region. Streams filled with mud and silt, resulting in poor water quality.

President Theodore Roosevelt created the National Forest Service to protect forests and watersheds from damage. Some of the first land bought was 2900 hectares (7200

The Monongahela forest is in demand for many reasons, including recreation, logging and mining jobs, and water supply. Balancing the impact from each use is a challenge for forest managers. In addition, air pollutants from sources outside the forest produce damaging acid rain in the forest.

Monterey Bay National Marine Sanctuary

The central coast of California supports a unique ecosystem known as the kelp forest. Kelp grows from the rocky seabed to the ocean surface. These forests of kelp are home to fish, sea otters, snails, sea urchins, crabs, and many other organisms.

Monterey Bay National Marine Sanctuary is one of 13 national marine sanctuaries. A marine sanctuary is like a national park in the ocean. Every national marine sanctuary protects ocean waters, the habitats found in them, and the local cultural history. Some activities, like dumping waste and drilling for oil, are not allowed in Monterey Bay National Marine Sanctuary. Other activities, like fishing and recreation, are permitted but regulated.

In the spring, water from deep underwater canyons flows up to the surface. The water is cold and contains lots of nutrients that phytoplankton need in order to grow. This upward movement of water is called upwelling. Upwelling results in tremendous productivity by phytoplankton, the base of the food pyramid in Monterey Bay National Marine Sanctuary.

The national marine sanctuary system was founded in 1972. Monterey Bay is the largest of the marine sanctuaries, covering 13,730 square kilometers (km) (5360 square miles). The sanctuary stretches from San Francisco to Santa Barbara, from the shoreline out into the ocean an average of 50 km (31 miles).

A primary concern for the kelp forest is the impact of fishing, kelp harvesting, and mariculture. Other issues include sea-otter abundance, jade collecting, and seabed disturbance.

Saguaro National Park

resting in the shade. Lizards and snakes might surprise you as they move from a place where they had been sunning to a cooler place.

Tucson, Arizona, is surrounded by the Sonoran Desert. In 1933 an extensive forest of saguaro cacti was preserved as a national monument by the federal government. The winter of 1937 was very cold, and some of the saguaros were damaged by frost. People thought that the freeze damage was a disease, and they worried that all saguaros were at risk. This worry led to the preservation of another forest of saguaros to the west of Tucson. In 1994 Saguaro National Monument became Saguaro National Park, preserving the habitat for the forests of the unusual cacti and all the populations that live in the desert ecosystem with them.

The primary issue in Saguaro National Park is damage to plants and animals by off-road vehicles. Expansion of cities into the desert is another concern. In the past an important issue was cattle grazing.

The southwestern United States includes California, Nevada, Arizona, Utah, New Mexico, and Texas. This region of the United States has many dry habitats, including deserts. The Sonoran Desert is in Arizona, part of California, and a bit of Mexico. This hot, dry desert is the most lush and diverse of the North American deserts. This ecosystem has many unique plants, including saguaro cacti, mesquite trees, ocotillos, and other succulent plants. During the day, the desert seems quiet, except for the occasional bird. A closer look might reveal rabbits and piglike peccaries

Yellowstone National Park

The most unusual feature of Yellowstone National Park is the geothermal activity. Magma close to the surface of Earth creates geothermal pools and geysers. Old Faithful is a geyser that throws superheated water more than 55 meters (180 feet) into the air every 76 to 100 minutes. Geothermal pools contain water heated by the same underground geothermal activity. The water in these pools is so hot that few organisms can live in them. They harbor several species of colorful cyanobacteria and algae, which turn the geothermal areas bright shades of yellow and orange.

In 1872 the U.S. Congress established Yellowstone National Park, the first park of its kind in the world. The park is in the Greater Yellowstone ecosystem, which also includes Grand Teton National Park and several national forests in Wyoming and Montana.

Yellowstone National Park includes several distinct ecosystems. There are unusual thermal pond ecosystems, freshwater lakes and streams, grasslands, and forests. The dominant habitat is taiga, or boreal forest. Taiga thrives in cold continental or subarctic climates. The dominant taiga plants are trees like pines and spruce.

Yellowstone National Park supports a diverse community of large wildlife, including elk, moose, deer, coyotes, and bears. Lucky visitors may see a wolf or one of the three species of large cats that inhabit the park.

A primary issue in Yellowstone National Park at this time is the reintroduction of wolves. Other issues are fire management and winter snowmobile use.

ADAPTATIONS

Imagine you are taking a vacation at Monterey Bay in California. Just offshore by the mouth of the bay is an extensive kelp forest. The long stemlike stipes of these huge algae reach 50 meters (164 feet) or more from the rocky seabed to the surface. Bobbing at the water's surface and diving down to the bottom are several members of a population of sea otters.

Slip into your one-person submarine and join the subsurface community. Fish swim among the kelp stipes, and jellyfish float on the current. Small kelp crabs, exactly the same color as the kelp, cling to the kelp fronds. As you descend to the bottom, you spot a lingcod resting between a couple of rocks, and spiny sea urchins are everywhere. Stuck tightly to the large rocks are abalones (ab•uh•LONE•ees), shellfish the size of small plates. If you are lucky, you might catch a glimpse of a harbor seal, a shark, or even a gray whale cruising through the area.

This is a typical central California marine nearshore ecosystem. The diversity and concentration of life are awesome.

The fun continues. The next week you visit Rocky Mountain National Park in Colorado. Three thousand meters (almost 10,000 feet) up in the Rockies you are surrounded by a different forest, this time pine, fir, and spruce. The trees are full of flying and climbing animals like jays, thrushes, woodpeckers, warblers, wasps, flies, and squirrels. In the understory are willows, wildflowers, and grasses. These

are visited by chipmunks, lizards, ants, bees, grasshoppers, crickets, and butterflies. If you sit quietly, you might see a deer, mountain sheep, or coyote pass through the area. This is a typical mountain-forest ecosystem, humming with a wide diversity of life.

Reflect for a moment. The organisms in the marine kelp forest and the organisms in the mountain forest are completely different. Not one sea otter lives in the Rocky Mountains, nor does a single chipmunk live in the kelp forest. Why?

The answer is that every organism has **adaptations** that make it possible for it to live in a particular environment. Adaptations are structures, characteristics, and behaviors that increase an organism's chances of surviving and reproducing in a particular environment. Kelp-forest organisms are fine-tuned for life in cold, turbulent seawater; mountain-forest organisms are adapted for life on land in the cool mountain air.

What adaptations do kelp-forest organisms have? The kelp-forest environment is cold, aquatic, salty, and turbulent. Anything that lives there has to be able to withstand chilling temperatures year-round, salt water, and tremendous tidal and storm surge. Let's look at tidal surge. Organisms need to have methods for staying put. Different organisms have different adaptations for staying in the ecosystem. The kelp have strong rootlike structures called holdfasts and limber, flexible stipes. The holdfast

grips rocks solidly, and the rest of the alga waves and flows in the tidal surge. The strong holdfast and flexible stipe are two **structural adaptations** the kelp has for staying in the ecosystem.

Sea otters wind themselves up in the floating fronds of the kelp to keep from being swept away. Young otters must learn this trick from the older otters. Anchoring in the kelp is a **behavioral adaptation** that keeps sea otters from being swept out of the ecosystem.

The abalone is a dome-shaped mollusk (related to snails) with a large foot that acts like a suction cup. When the surge is strong, the powerful suction and streamlined shape of the shell allow the abalone to stay attached to the rocks. Are these structural or behavioral adaptations? The shape of the shell is structural, as is the large suction

foot, but the act of clamping down in response to increased surge could be considered a behavioral adaptation.

Other examples of adaptations in the kelp forest include thick fur (sea otter) or layers of insulating fat (seals and whales) to protect against the chilling effects of cold ocean water. Warmblooded animals without some form of insulation are not adapted for life in the kelp forest. Spines (sea urchin), hard protective coverings (crabs and snails), fins for fast swimming (fish), and protective coloration (kelp crabs) are all structural adaptations that increase the organism's chance of survival. Nocturnal hunting (octopus), egg guarding (lingcod), and kelp anchoring (sea otter) are all behavioral adaptations for survival.

A few adaptations are shared by *all* the organisms in the kelp-forest ecosystem. All are adapted to live in a salt environment. All are adapted to live in cold water. All are adapted to withstand the rigors of tidal surge. They don't have the *same* adaptations for dealing with these important environmental factors, but they all have adaptations. By contrast, none of the organisms in the mountain-forest ecosystem has an adaptation for even one of the major kelp-forest factors.

What are some of the adaptations exhibited by organisms in the mountain forest? What adaptations do the trees have for survival and reproduction? What adaptations do the birds and squirrels have? What about the wildflowers and grass? The woodpecker's beak, the butterfly's brightly colored wings, the buck's antlers, the chipmunk's cheek pouches, the cricket's chirp, the grasshopper's long legs—they all are adaptations that help each organism survive in its ecosystem.

Let's change locations again. In this ecosystem we find the barrel cactus, an unlikely-looking plant if we are still thinking about the forest ecosystem. The cactus doesn't have the structures we are accustomed to seeing. No trunk or stem really, no branches, and no apparent leaves. It does have roots and flowers, and forms seeds, so it is a plant, but a strange one.

Think about the desert where a plant like this lives. What are the usual conditions in the desert? It is frequently windy, intensely sunny, often hot, and almost always dry. An organism would thrive in this environment if it had adaptations for dealing with these extremes of climate. Leaves are necessary for most plants as sites for photosynthesis. But broad, flat leaves are a liability if it is windy, intensely sunny, and hot. A good adaptation for wind and intense light would be fewer and smaller leaves.

The barrel cactus has taken this adaptation to the limit. We can imagine that over millions of years of variation in the cactus population, the individuals with the smallest and fewest leaves were the ones that survived best. The successful descendants of those earlier plants have just

the bare remains of leaves. The spines seen on the barrel cactus today are its tough, highly adapted leaves.

The barrel cactus has adaptations for dealing with the scarcity of water as well. Just about the whole body of the cactus, which is a highly adapted stem, is devoted to storing water. The spongy tissue inside the cactus holds many liters of water, which can be used by the cactus to stay alive during extended dry spells. And all those spines protect all that water from other organisms that would like to have it. The spines are adaptations for protecting the stored water.

Some organisms are able to live in a number of different environments. The coyote is one example. Coyotes live in high mountains, marshes, prairies, forests, deserts, and towns. They have long legs for running, a keen sense of smell, good eyesight, fur to protect them from the elements, and the ability to eat almost anything. In addition, coyotes travel in packs, dig dens to protect their young, and hunt day and night. These structures and behaviors are general adaptations that give coyotes flexibility to survive in many different terrestrial ecosystems. Can you think of other organisms with general adaptations that give them the flexibility to survive in many ecosystems? *Homo sapiens* is one. Think about all the places people live. What is it about humans that makes them so flexible? What adaptations do we have that make it possible for us to live on land, underwater,

in space, and all over the planet's surface?

Adaptations are the keys to survival. Every organism alive on Earth has adaptations that allow it to live and reproduce in its environment. Organisms are born with their adaptations. If conditions change, individual organisms can't change their adaptations to cope with the new conditions. Those organisms that were born with adaptations to deal with the new conditions are the ones that will survive, reproduce, and continue the population.

THINK QUESTIONS

1. What are structural adaptations? Give three examples.

2. What are behavioral adaptations? Give three examples.

3. Frogs live in pond environments. List three frog adaptations and describe how they help the frog survive.

4. A trout stream runs through a meadow where the stream bottom is dark and there are lots of shadows. Then the stream flows down through some rapids where the water splashes over and around light-colored rocks. The trout have dark backs in the meadow and light backs in the rapids. How do you explain the color variation in the trout population?

Organisms produce offspring. The offspring usually look pretty much like their parents. Catfish offspring are catfish. Tomato-plant offspring are tomato plants. Monarch-butterfly offspring are monarch butterflies. Clearly some kind of information was passed from the parents to the next generation. Something inherited by the offspring from its parents carried the message for how to develop just like Mom and Dad.

This fact was known for thousands of years, based on observation. Distinct facial features were passed from generation to generation. The mating of large, strong horses usually produced large, strong offspring. Goats with long, silky hair usually produced more of their kind. People used common sense to select and breed plants and animals with desirable characteristics. The offspring inherited something that produced the desired characteristics. But what was inherited? It was a mystery.

Early Research into Heredity

Gregor Mendel was born in 1822 in a poor farming community in what is now the Czech Republic in central Europe. He was a bright student, but his family didn't have enough money to send Gregor to the university. In order to continue his studies he joined a monastery. There he made a life of teaching and researching the question of heredity. By conducting careful experiments over many years, Mendel made landmark discoveries in heredity and established a new science: genetics.

Mendel was a keen observer of nature. He noticed that the common garden pea, *Pisum sativum,* had a significant amount of variation from plant to plant. Some plants produced green seeds, others yellow seeds; some plants produced red flowers, others white flowers; some plants were tall and others short; and so on. Mendel decided to see what he could find out about the distribution of the characteristics.

The experiments Mendel designed proceeded in three phases. First, he raised several generations of pea plants, making sure that the plants self-pollinated. That means that the flowers on every plant were fertilized only with pollen from the same plant. No flower was fertilized with pollen from another plant. In this way Mendel obtained pure breeding strains of plants for a trait, such as plant height. Tall plants produced only offspring that became tall plants when mature. Another strain produced only offspring that were short.

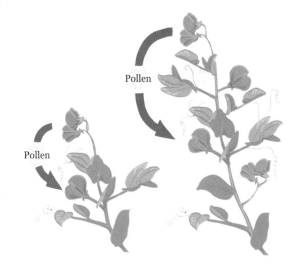

Pollen

Pollen

Next Mendel allowed his pure breeding strains to crossbreed. Pollen from tall plants was placed on the flowers of short plants, and pollen from short plants was placed on the flowers of tall plants.

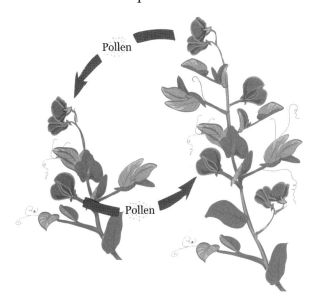

The plants that he allowed to cross-pollinate Mendel called the parents. He identified them as the **P generation.** The first generation of offspring from the parents Mendel called the first filial generation (filial means sons and daughters). He identified them as the F_1 **generation.**

The third phase of Mendel's experiment involved letting the F_1 generation self-pollinate. The offspring of the self-pollination was the F_2 generation.

Mendel's Results

Mendel crossed his tall-plant and short-plant parents to produce an F_1 generation. When the F_1 generation matured, all the F_1 plants were tall. There were no short plants in the F_1 generation.

The tall F_1 plants were isolated from one another. Pollen from the flowers on each plant was used to pollinate the other flowers

on the same plant. Some of the offspring, the F_2 generation, were tall plants and some were short plants!

Mendel had to make sense out of the results. The original parents were pure breeders for height (either tall or short). The F_1 generation produced by crossing a short plant and a tall plant were all tall. The tall F_1 generation produced both tall and short mature plants when allowed to self-pollinate. Below is a diagram of Mendel's results. (Each square represents a plant and indicates its height, with gray being tall.)

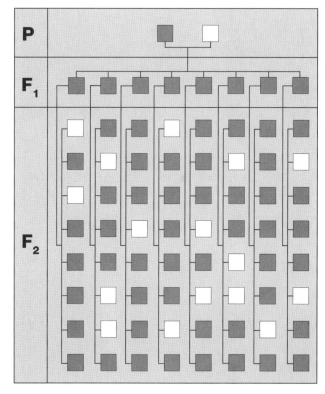

When Mendel counted the number of tall and short plants in the F_2 generation, he found one short plant for every three tall plants. The ratio of tall to short plants was 3:1. What could cause one out of four plants to be short, when short plants were completely absent from the F_1 generation?

Mendel's hypothesis revolutionized our understanding of heredity. This is what he reasoned.

Height is a **feature** of pea plants. The feature has two **traits,** tall and short. Offspring inherit something (Mendel didn't know what) that determines the trait from two sources, one from the male and one from the female. Mendel called the inherited something a factor.

The pea plants, which go through sexual reproduction, must get one height factor from the pollen (male) and one height factor from the ovule (female). As a result, every plant has *two* factors for every feature, including the feature of height.

During pollination, *one or the other* of the two factors is passed to the offspring. When Mendel crossed pure tall plants and pure short plants (pollen from tall plants onto short-plant flowers, and pollen from short plants onto tall-plant flowers), only the tall trait appeared in the F$_1$ generation. All the F$_1$ offspring had to have one tall-plant factor and one short-plant factor, but only the tall trait appeared. Mendel called the tall-plant factor **dominant.**

The short-plant factor, which Mendel reasoned had to be there, was **recessive.** The recessive factor could appear only when the offspring inherited the recessive factor from both parents. In that case the short-plant trait could appear because there would be no dominant factor.

Punnett Squares

Mendel's hypothesis explained how traits could disappear in one generation and reappear in the next. His ideas explained observation in the real world. Furthermore, he could predict the number of offspring that would have a dominant trait and a recessive trait.

The method we use today to predict the traits of offspring is the Punnett square. Reginald Punnett, an early 20th-century heredity scientist, introduced it. His method uses a simple two-coordinate system to show the probability of traits in offspring. This is how it works.

The feature of plant height is represented by the letter *t* for tallness. An uppercase letter *(T)* refers to the dominant factor and a lowercase letter *(t)* refers to the recessive factor.

Mendel's true-breeding tall pea plants had to have two dominant factors (TT). The true-breeding short pea plants had to have two recessive factors (tt). A Punnett square set up to represent Mendel's cross of the two true-breeding pea plants looks like this.

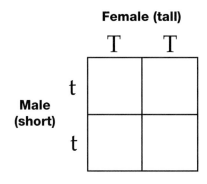

Filling in the squares with the factors produces four possible offspring, and they all have the same pair of factors (Tt). Because all the offspring have a dominant factor, they all have the tall-plant trait.

Female (tall)

	T	T
t	T t	T t
t	T t	T t

Male (short)

Let's see what happened when Mendel self-pollinated the F$_1$ generation of tall plants. The male (pollen) and female (ovule) factors would have to both be Tt.

Female (tall)

	T	t
T	T T	T t
t	T t	t t

Male (tall)

Of the four possible offspring, three have the dominant factor, one TT and two Tt. The fourth has two recessive factors (tt), however, producing a short-plant offspring.

Mendel's experiments, completed in 1865, uncovered two principles that proved to be huge breakthroughs in the science of heredity.

1) Two factors determine traits. One factor comes from each parent. 2) Factors can be dominant or recessive. Recessive factors can be present in an organism even when no trait confirms that they are there.

Understanding the Heredity Factor

Mendel's discoveries were so advanced that other scientists could not understand or accept their meaning. His work drifted out of the mainstream of science for 35 years. During those years, however, microbiology was making strides. Better microscopes and advanced methods for preparing samples provided better understanding of the structure of the nucleus of the cell.

In about 1875 chromosomes were observed inside the nucleus of a cell. They looked like little Xs and hot dogs of different sizes. A few years later scientists observed that, when a cell prepared to divide, the chromosomes first duplicated themselves. After division, both new cells had identical sets of chromosomes. In fact, every cell in an organism had exactly the same set of chromosomes. And all members of the same species have the same number of chromosomes.

In 1902 Walter S. Sutton observed that the chromosomes in a nucleus could be sorted into almost identical pairs. The grasshopper that Sutton was studying had 18 chromosomes. When he looked closely, he could see nine almost identical pairs. He named the two members of a pair **homologues** from a Greek word homologos.

The nucleus of a make-believe animal, the larkey, has eight chromosomes, organized in four homologous pairs. A larkey nucleus looks like this.

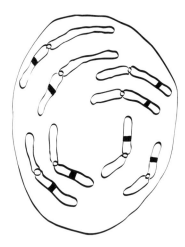

Further investigation revealed that, when cells divided to produce sex cells (sperm and eggs), the homologous chromosomes separated and went into separate cells. The sperm and egg cells had just one set of chromosomes, not two.

This discovery was remarkable. Sutton realized that, if the factors of heredity (now known as genes) were located on chromosomes, the two genes that determined a trait could be on homologous chromosomes. One form of the gene (perhaps a dominant gene) could be on one of the homologues and another form of the gene (maybe a recessive gene) could be on the other homologue. When a cell divided to form two sperm cells in a male (or egg cells in a female), the two homologous chromosomes separated. One chromosome in the pair went into one sperm cell, and the other chromosome went into the other sperm cell. Sperm cells and egg cells have just one set of chromosomes in their nuclei.

During sexual reproduction a sperm cell and an egg cell fuse. This is called

fertilization. The set of chromosomes from the sperm and the set of chromosomes from the egg are united to form a *new* set of homologous chromosomes. The chromosomes are again paired, but the homologous pairs are different than those of either parent. The whole mechanism of heredity fell into place.

During the next decade it was confirmed that genes were indeed carried on chromosomes, and the two forms of the gene on homologous (paired) chromosomes could be identical (both dominant or both recessive), or they could be different (one of each).

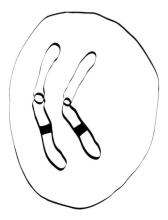

The vocabulary was refined to make communication more precise. The form of a gene on a single chromosome was an **allele,** and the two interacting alleles on homologous chromosomes constituted a **gene.**

What Is a Gene?

Understanding the gene came from understanding the structure and function of one remarkable molecule, deoxyribonucleic acid, better known simply as **DNA.** It was first extracted from the nucleus of cells in 1869. Because it is slightly acidic, it was named nucleic acid.

During the 1930s the great biochemist Phoebus A. Levene analyzed the DNA molecule and found that it was composed of several subunits—a phosphate unit, a sugar, and four nitrogenous bases. Levene got the analysis of the components of DNA exactly right. What he failed to understand was how they all fit together in a long repeating chain. This error in determining how the phosphates, sugars, and bases went together stalled further understanding of the role of DNA.

In the early 1950s two young scientists met at the Cavendish Laboratory in Cambridge, England. James D. Watson was a molecular biologist, and Francis H. Crick was a physicist. They proceeded to study what was known about the structure of the DNA molecule. The chemistry of the parts was known, as was the ratio of the four nitrogenous bases. The ratio of guanine to cytosine was 1:1; the ratio of thymine to adenine was also 1:1. Evidence from other sources indicated that the long molecule might be wound up in a spiral.

The data suggested that the DNA molecule had a long backbone of phosphates and sugars with the nitrogenous bases sticking out to the sides.

The single-stranded molecule didn't, however, form a coil. After exhaustive further study of their own data, and using crystallography structured data from another scientist, Rosalind Franklin, Watson and Crick deduced that the DNA molecule was not a single long strand wound into a spiral, but a double strand. Two backbones of phosphates and sugars with bases sticking out were running right next to each other.

Watson and Crick began to build models. The bases acted as rungs between the two phosphate-sugar rails. The model began to look like a twisted ladder. When they realized that the base cytosine could form a rung only with guanine, and adenine could form a rung only with thymine, they had it! The structure of the magnificent molecule of heredity had been discovered, and it fit all the criteria for the central player in the story of heredity.

The DNA molecule is extremely long. The DNA in one of your cells might be 150 centimeters (59 inches) long if it were stretched out straight. To fit in the nucleus of a cell, the molecule is twisted into a spiral, called a helix. That helix is twisted into a second helix, and that helix is wound again, and so forth, until it forms the chromosome. That's what a chromosome is—a DNA molecule, twisted into a compact structure.

So where is the gene? A section of DNA—a sequence of bases—codes for a feature. A sequence may be short or long, and any sequence of bases along the length of the millions of bases might be a gene. Corresponding sequences on two homologous chromosomes—the two alleles—constitute a gene. Genes direct the manufacture of proteins, which go out into the body to make things happen.

Genotype and Phenotype

Genes determine features. The two alleles a larkey has for eye color interact to produce the trait of gray eyes or red eyes. If the larkey has two dominant alleles (EE) for eye color, the eyes will be red. If the larkey has one dominant and one recessive allele (Ee) for eye color, the eyes will still be red. Only if the larkey has two recessive alleles (ee) will the eyes be gray. The two alleles for a feature on a larkey's paired homologous chromosomes is its **genotype** for that feature.

If a gene is represented by two identical alleles, the larkey is **homozygous** for that feature. A homozygous genotype can be either homozygous dominant (EE) or homozygous recessive (ee). If a gene is represented by one dominant and one recessive allele, the larkey is **heterozygous** for eye color. The heterozygous genotype for larkey eye color is Ee.

Of course, larkeys have lots of other features. Larkeys have a pair of alleles that constitutes the gene for every feature. The larkey's complete set of paired alleles is its overall genotype. Here are three of the many possible genotypes for the four larkey genes being investigated in this course.

A a		A a		A A
E E		E E		E e
f f		f f		f f
T T		T t		T T

Genes code for features. Particular pairs of alleles—genes—will determine what traits an organism has. If a larkey has even one dominant gene for leg length in its genotype (that is, it is either homozygous dominant or heterozygous), the larkey will have short legs. How an organism looks as a result of its genotype is its **phenotype.** Because homozygous dominant (EE) and heterozygous (Ee) genotypes both produce

phenotypes with the dominant trait, genotype is not always apparent from the phenotype. The three larkey genotypes used as examples earlier actually produce identical phenotypes.

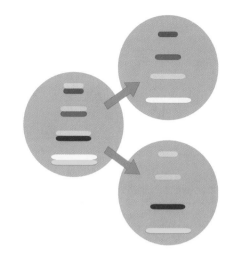

How Do Genes Get Mixed Up from Generation to Generation?

Organisms grow and maintain themselves by producing new cells. New cells result from cell division—a cell simply divides into two cells exactly like the old cell. Just like that, there are twice as many cells.

In preparation for cell division, a cell first produces a complete set of new chromosomes. When cell division happens, one set of chromosomes goes into each new cell. In the larkey case, the cell briefly has 16 chromosomes. Then it divides, and each cell gets one set of eight chromosomes (four homologous pairs). Both cells are exactly the same. This process is known as mitosis.

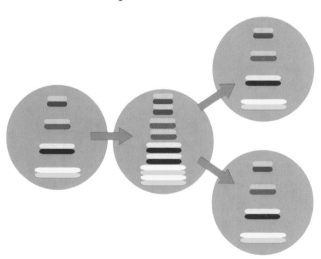

The story is different when cells divide to form sperm and egg cells as they do in organisms that reproduce sexually. After dividing once to form two identical cells, each of those divides again. But the chromosomes do not duplicate before

division. Instead, one member of each chromosome pair goes into one of the new cells, and the other chromosome goes into the other cell. These are the sperm cells or the egg cells. As a result each larkey sperm or egg cell has only four chromosomes instead of the usual eight. This process is known as meiosis.

During fertilization, the sperm cell nucleus fuses with the nucleus of the egg cell. At that time the four sperm chromosomes match up with the four egg chromosomes. The set of chromosomes is again complete—four pairs of homologous chromosomes. That creates the genotype for a new larkey.

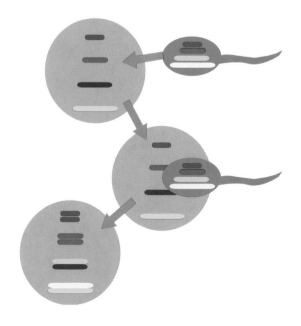

The genotype is usually different from the genotype of the mother and different from the genotype of the father. Half the alleles for the genes come from the father and the other half come from the mother. This explains how offspring can be similar to their parents and yet unique.

Mapping the Human Genome

You have a genome. A genome is an organism's complete set of DNA, the incredibly long molecule that chromosomes are made of. If you could take apart all your chromosomes and look at the long string of base pairs along the double helix, you would be looking at your genome.

Every organism has a unique genome. At the same time, however, every organism has almost exactly the same genome as all other members of its species. More than 99.9% of your nucleotide bases (adenine, thymine, cytosine, and guanine) are exactly the same as your mother's, your brother's, your best friend's, your favorite rock star's, Nelson Mandela's, and everyone else's on Earth in Europe, Asia, Africa, the South Pacific islands, and everywhere else. There is a human genome.

In 1990 the United States Human Genome Project was launched. It was coordinated by the U.S. Department of Energy and the National Institutes of Health. The goal of the project was simple—identify the sequence of all the base pairs in all the chromosomes in the human genome. Accomplishing the goal is extremely difficult. There are about 3.1647 billion (3,164,700,000) base pairs. The strings of DNA in the individual chromosomes range

from 50 million to more than 250 million bases. The job was huge.

The project was originally expected to take 15 years, but rapidly advancing laboratory technologies shortened the time by a couple of years. The first working draft of the sequence of bases in the human genome was released in June 2000. Genetics researchers had some interesting results.

Humans have a surprisingly small number of genes. Early in the project, scientists estimated that the human genome had between 80,000 and 120,000 genes. As the project draws close to completion, the number has been revised to 30,000 to 35,000 genes. Genes, sometimes called locations on chromosomes, are sequences of bases along the genome that code for the synthesis of a protein. If a particular sequence of bases were to move from location A to location B on the DNA molecule, the gene would be unchanged, and it would still code for a protein.

Gene sequences can be as small as a few hundred or as big as 2.4 million base pairs. Average gene size is about 3000 base pairs.

Most human DNA doesn't code for proteins; that is, it has no apparent genetic function. It is sometimes called junk DNA. The inactive junk DNA areas seem to act as spacers between the active genetic regions of the genome. Only about 2% of the genome is genetically active.

But what a marvelous job that 2% does! The human genome, which is now written down in its simple four-letter alphabet, G, C, A, and T, is available for all to read. The amazing story it tells is how to make a human being. The directions for building and maintaining the most complex system known is written in living code and tucked away in the nucleus of every cell in your body.

Conclusion

In 1865 Mendel announced to the scientific community that organisms pass units of information to their offspring during reproduction. This inheritance allows the offspring to develop just like their parents. He didn't know what the units were, but he understood how they acted. We now know that Mendel discovered the existence of genes and described how they work.

For a number of reasons Mendel's discoveries were forgotten for 35 years. During those years the field of biochemistry was making advances. The DNA molecule was extracted from the chromosomes in the nucleus, and it was analyzed. The phosphate group, sugar molecules, and four nucleotide bases were identified. But how, or even if, this huge mass of chemical units contributed to heredity was still a mystery.

In 1953 Watson and Crick announced their model for the structure of the DNA molecule—the double helix. Overnight the mechanism of inheritance was clarified. Sequences of bases along the huge DNA molecule acted as the instruction manual for synthesizing proteins that make life possible. And the mechanisms for replicating the DNA molecule further clarified the story of inheritance. The mystery was solved in the mid-20th century.

The quest for understanding continues, however. At the close of the 20th century, not even 150 years after Mendel puzzled over his pea plants, the scientific community proposed a bold project to read the entire human genome—to identify every single base pair of the DNA of a human being. The 15-year project is drawing to a close, and the genome is available for all to read.

What stories will the genome tell? Is the cure for cancer written there? Can the information coded in the DNA be used to make proteins to repair body structures or manufacture specific drugs? Can the story written in the genome describe how we are related to all other life-forms? As usual, the closing of one chapter in science marks the opening of another. The completion of the sequencing of the human genome opens the doors to new science, with applications that we can only dream about today.

A Larkey Yammer

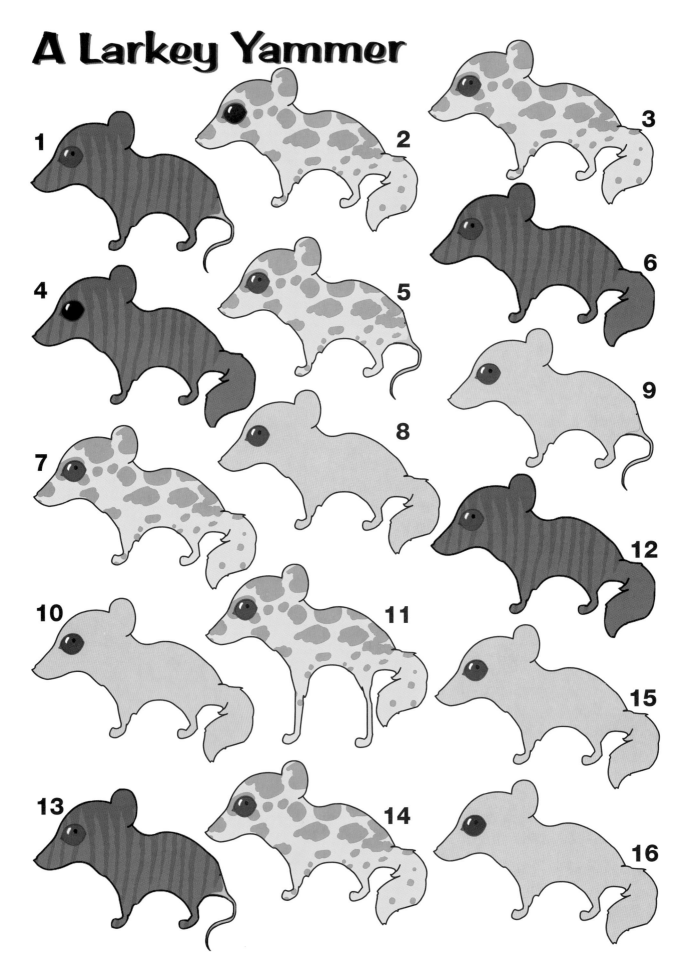

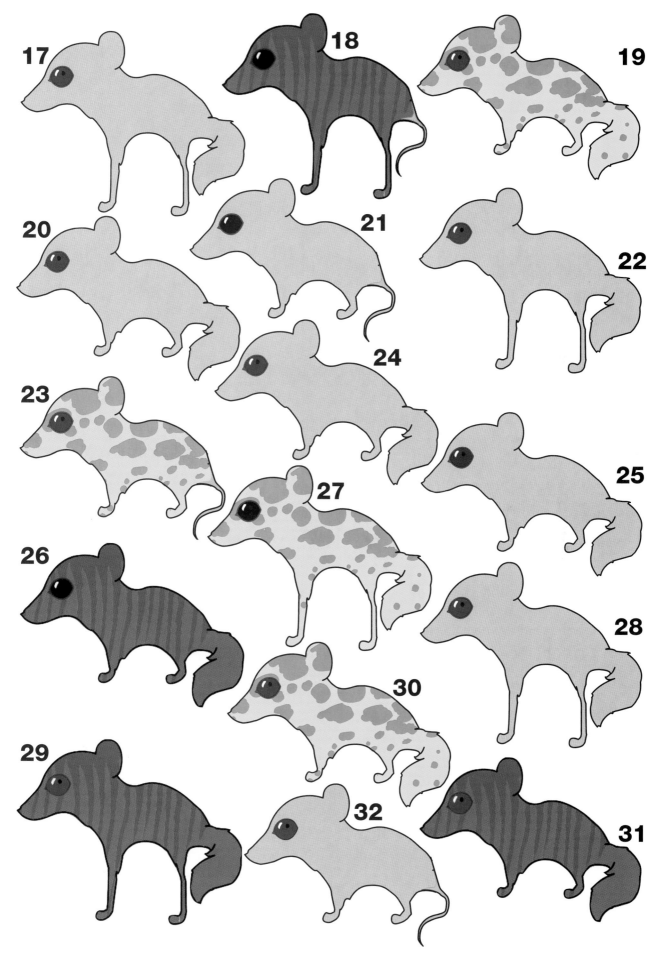

Natural and Unnatural Selection

If you rounded up all the dogs in your neighborhood or town, you would see an amazing diversity of critters. There could be little yappy Yorkshire terriers and skinny Chihuahuas, low-slung dachshunds and corgies, silky collies, fleet, long-legged greyhounds, grumpy-looking pugs, water-loving black Labrador retrievers, huge mastiffs and Great Danes, and dangerous, brooding pit bulls. Where did they all come from? Why do they look so different?

Suppose you wanted a dog to chase badgers out of their underground burrows. The dog would have to be short-legged and armed with an impressive set of teeth. If there was a dog with shorter legs in the community, you would make sure it produced offspring. You would breed the offspring with the shortest legs and then the short-legged offspring again. In a few generations you might have a bunch of short-legged dogs.

There is now substantial scientific evidence that the domestic dog, in all its interesting sizes, shapes, and temperaments, evolved from the wolf. There is evidence that one strain of wolves may have started hanging around human settlements as much as 135,000 years ago. These wolves would not qualify as lapdogs at that time, but they were comfortable around humans.

The modern history of dogs starts about 2500 years ago. Several kinds of dogs played roles in human culture, including hunting, defending livestock, and carrying loads. Today there are about 400 distinct breeds of dog, most of which have come into existence in the last 200 years. Why so many? Selective breeding.

Among them you would look for the ones that were apt to plunge into burrows to take on the ill-tempered badger. You would breed those with the right shape and attitude about badgers to produce offspring with those characteristics. This might have been the process that produced the dachshund.

Humans decide which qualities in a dog are desirable, find individuals in the population that have those traits, and breed them to reinforce those traits. Individuals that breed are selected by humans, not by natural processes. This **selective breeding** of organisms to produce offspring with desirable traits is also known as artificial selection.

Selective breeding has been used to produce the hundreds of breeds of dogs, breeds of horses for different uses, breeds of cats, chickens, hogs, dairy cows, wheat, rice, tomatoes, corn, and so on. The desirable traits can vary widely, ranging from aesthetic appeal, rate of maturity, season of production, yield of product, and so on. Humans are very good at manipulating the genotype of organisms by selective breeding to produce phenotypes that satisfy some need.

Natural Selection

In nature, nobody consciously selects the individuals that will breed to produce offspring. Even so, some individuals are selected to reproduce, and some are not. Who or what does the selecting? The answer is the environment.

Life is usually a struggle. Limited resources, food supply, predators, weather, and access to mates all put pressure on organisms all the time. Individuals in populations that are adapted to the environment in which they live have ways to respond to the pressures. Because all populations have variation, some individuals will be better at responding to each source of pressure than others. Some will be better at getting resources, others will be more efficient at finding food, and some will have better strategies for avoiding the effects of weather.

Some individuals will be better at finding a mate and breeding, thereby passing heritable traits on to more offspring. Bright coloration, large antlers or horns, the ability to defend a territory, or the performance of courtship displays are examples of structures or behaviors (features) that make the males of some species more attractive to mates. When the adaptive feature is enhanced by female choice, it is referred to as sexual selection. This is a selective pressure that influences populations.

A change in the environment might apply new pressure on a population. The change might be a new predator, a drought, increased competition for a favored food source, a cold snap, or any of a thousand other things. Some individuals in the population will be, because of variation, better able to withstand the increased pressure. They will be more likely to reproduce and leave offspring with the traits that allow them to deal with the pressure. If the pressure persists, in time the population will have different traits than it did before the pressure.

An interesting study was conducted on a guppy species in South America. A biologist, John Endler, found that guppy populations along one river were isolated from one another by waterfalls. The guppies in pools higher up in the river had few predators. In these pools the males were brightly decorated with different numbers and sizes of colored spots. The female guppies were plain. But the males in pools downriver where there were large numbers of predators were not colorfully decorated, but rather had small spots and few of them. Endler planned a study to investigate the pressures on the guppies. He moved a population of 200 guppies from downriver (with aggressive predators) to a pool high up that had no severe predators and no guppies. In 2 years (14 generations) the males in the population, which had been plain, were brightly colored with big spots.

The biologist reasoned that this might be evidence of the interaction of two selective pressures. Female guppies prefer to mate with brightly decorated males. But these brightly colored males were more easily seen by predators and removed from the population in the lower pools. Without the pressure of the predators in the higher pools, after several generations the males displayed bright colors and large spots. Variation in male guppy coloration existed in the population and the selective pressure acted on it. The females' choice of the more brightly colored males was a selective pressure and resulted in brightly colored male offspring but only when the pressure of predators was gone.

Who selected the individuals that would reproduce to pass along their traits? Ultimately, the environment! When the biotic and/or abiotic environment changes, it places selective pressure on the organisms in that environment. Those individuals that are adapted to withstand the pressure will reproduce; those who can't take the pressure will not reproduce. The genes of the survivors are passed to the next generation, and with them the adaptive traits. That's **natural selection.**

When Charles Darwin visited the Galápagos Islands in the 1830s, he noticed a diverse community of finches. He reported that each kind had a modified beak that served its lifestyle. After extensive study of Darwin's finches, it is now understood that all 13 different finches on the Galápagos Islands are descendants of one species of finch that arrived on the islands thousands of years ago. How did one species of finch evolve into 13?

Darwin imagined that a flock of finches was blown out to sea during a big storm in the distant past. The flock landed, not on the South American mainland, but on a ragtag collection of relatively new volcanic islands. The islands had none of the familiar food sources and nesting sites. The small finch population was in a radically different environment.

The islands had a number of potential food sources. Seeds of several sizes grew on grasses, shrubs, and trees. The finches undoubtedly tried to use all sources as food.

Because there was variation in beak size in the population, individuals with larger, stronger beaks could crack large seeds; individuals with smaller beaks could not. On the other hand, finches with smaller beaks could easily gather large numbers of small seeds; individuals with large beaks could not.

Finches that fed on large seeds mated with one another and produced offspring with large, strong beaks. The larger, stronger beaks allowed the large-beaked finches to survive when large seeds were plentiful. Isolation from the other finches, due in this case to food preference, produced, over a long time, enough differences between the large-beaked finches and the small-beaked finches that they could no longer mate to produce offspring. They had evolved into two new species.

This same process of eating different foods based on beak size and shape produced other divisions in the populations. Over time, divisions into subpopulations resulted in 13 different species.

What would happen to the large-beaked finches if the large nuts became very scarce? They would have to turn to other food sources. Within the population of large-beaked finches would be individuals with smaller beaks. The small beaks would give those individuals a feeding advantage. They would survive and reproduce. Their offspring would inherit the trait of smaller beaks. The selective pressure of food source would shift the population toward smaller beaks. The environment rules. Those with phenotypes that allow them to survive pass their genes to the next generation, and with them their traits.

Darwin's Finches Today

Recently the environment on the Galápagos Islands has changed. This time it was caused by a fly. The larval stage of the fly is parasitic on finch nestlings. The larvae, also called maggots, burrow into a chick's body, weakening and sometimes killing the host finch. The parasitic fly larvae are recent arrivals in the Galápagos Islands, probably accidently transported by ship or plane from North America.

The pressure imposed by the parasite is new, so it is not clear whether individuals in the finch populations have adaptations for defending against the deadly maggot or whether the invader will exterminate one or more of the populations of Darwin's finches.

The ancient struggle for survival continues. Because the environment is constantly changing, the organisms that survive and thrive are constantly changing as well. When the change in the environment is so fast or so extreme that no individuals in a population survive, the entire population will die, a condition called **extinction.** Extinction is the condition of most of the life-forms that lived on Earth at one time. The 10–30 million species living on Earth now are a tiny fraction of the total number of species that once graced the planet. And they came into being, had their day in the Sun, and died off, all as a result of natural selection.

Where Are the Checkers and Spreads?

By Melinda S. LaBranche

 CORNELL LAB *of* ORNITHOLOGY

Project PigeonWatch participants are helping us answer the question: How are various pigeon color morphs distributed across the continent? In Europe dark-colored pigeons are more common in the cool, northern latitudes. Perhaps the dark colors help the pigeons to absorb more heat from the Sun. Is this the same pattern we see in the United States?

Project PigeonWatch participants record the number of pigeons in each of the seven different color morphs as well as the total

pigments in their feathers. All flocks in this data set had at least one melanic pigeon, so I examined the distributions of melanics by latitude and longitude for 110 pigeon flocks across the United States.

Flocks at high latitudes (northern) and western longitudes had higher proportions of melanic pigeons than flocks in the southern and eastern United States. The distribution of melanic pigeons can be seen below. The size of the red circles indicates the relative proportions of melanic birds in a flock—notice that the largest circles (and thus the largest proportions of melanic pigeons) are in the Northeast and in California.

Color Morphs
· 0 – 0.2
· 0.2 – 0.4
● 0.4 – 0.6
● 0.6 – 0.8
● 0.8 – 1.0

Because white feathers reflect some of the Sun's heat, you might expect to have more white pigeons in the South. I examined the proportions of white pigeons in these 110 flocks and did not find any significant effect of latitude or longitude.

Feral pigeon flocks in northern and western cities had higher proportions of melanic pigeons than flocks in the southern and eastern United States.

number of pigeons in every flock they observe. From these data, we calculate the proportions of each color morph and examine the distributions of those proportions over large areas. Checker and spread color morphs are called melanic because of the black-colored melanin

We might conclude that a pigeon's home city and its color will affect how well it survives and reproduces. Over time, dark pigeons in the North may have more offspring than dark pigeons in the South, causing northern flocks to have higher proportions of melanic pigeons than southern flocks. But we have to make this conclusion with caution, because this sample is very small. As indicated on the map, many cities and states are not represented in the data. Even within each city represented, we do not have data from all the flocks. We need data from more flocks and from more cities and states. Sign up for Project PigeonWatch and help us unravel this color-morph mystery.

Project PigeonWatch is an international research project that involves people of all ages and locations in a real scientific endeavor. People participate by counting pigeons and recording courtship behaviors observed in their neighborhood pigeon flocks. Participants send their data to the Cornell Lab of Ornithology, where scientists compile the information and use it to examine two questions of scientific interest:

1. Why do city pigeons exist in so many colors (morphs)?

2. What color mate does a pigeon choose?

To join PigeonWatch, contact http://birds.cornell.edu/ppw/.

Pigeon morph illustrations by Julie Zickefoose

EARTHWORM

Genus: **Lumbricus**

Species: **sp.**

Size: Up to 25 cm (10 in.) long

Range: Worldwide except polar regions

Natural History: Earthworms live in the upper layers of the soil but will tunnel as deep as 2 m (6.5 ft.) if conditions are too dry or too cool. They prefer light, loamy soils to those high in clay and sand. Temperatures of about 13ºC (60ºF) are ideal.

Food: Earthworms eat decaying organic material (detritus). They decompose the organic material and return nutrients to the ground.

Predator: Birds, frogs, toads, salamanders, lizards, shrews, minks, raccoons, and turtles

Shelter: Tunnels in the upper layers of soil

Reproduction: Earthworms are hermaphroditic (have both male and female sex organs). The mating pair fertilize each other. An enlarged ring produces a cocoon. The cocoon is left behind in the soil when the pair separate. Tiny earthworms emerge in 2–4 weeks.

Abiotic Impact: The movement of earthworms helps break up and loosen soils.

Human Impact: Humans add earthworms to gardens to improve the quality of the soil. They also use earthworms as fish bait.

AQUATIC SNAIL

Genus: **Planorbis**

Species: **sp.**

Size: Shell up to 3 cm (1.2 in.) diameter

Range: Temperate and tropical freshwater ponds

Natural History: Snails have a hard, spiraled shell. It gets bigger toward the opening as the snail grows. The muscular part that protrudes from the shell is the foot. The snail scrapes algae from the surfaces over which it travels.

Food: Algae, aquatic plants, and detritus (decaying organic material). Snails eat whatever food is left over by fish.

Predator: Large fish, birds, waterfowl, shrews, and turtles

Shelter: If threatened, an aquatic snail pulls into its shell and closes the opening with an operculum.

Reproduction: Snails have both male and female sex organs (they are hermaphroditic). During mating, two snails exchange sperm and both may lay eggs. Eggs are laid on plants in a jelly capsule. The eggs hatch into tiny larvae that swim freely until they begin to grow a shell. The shell weighs them down, and they begin a life of crawling along the water bottom.

Abiotic Impact: Absence of calcium in the environment causes weak shells.

Human Impact: Aquatic snails are kept in home aquariums to reduce the amount of algae and detritus.

ISOPOD

Genus: *Oniscus*

Species: *sp.*

Size: 1.8 cm (0.7 in.)

Range: Temperate and tropical regions

Natural History: Isopods are found in dark, damp areas, especially under rocks and in leaf litter.

Food: Detritus (decaying organic material), fruit, fungi, and young plants

Predator: Birds, frogs, lizards, turtles, and salamanders

Shelter: Under rocks and logs, and in leaf litter. Some species roll into a tight ball if threatened.

Reproduction: Females carry eggs until they hatch.

Abiotic Impact: Must stay moist

Human Impact: Minor agricultural pest

GUPPY

Genus: *Poecilia*

Species: *reticulata*

Size: 3 cm (1.2 in.) long

Range: Freshwater streams of Central America and northern South America

Natural History: Guppies are small fish that bear live young. Females are usually beige or silver gray. Males are smaller and have longer, flowing tails.

Food: These omnivores eat detritus (decaying organic material), algae, smaller fish, and plants.

Predator: Larger fish and adults eat unprotected young.

Shelter: Females and young hide in vegetation. Guppies can live in an unheated aquarium unless room temperatures go below 15°C (65°F) or above 29.5°C (85°F).

Reproduction: Guppies breed easily in home aquariums. Females can store sperm and may produce several broods from one mating. Babies are born live and must find shelter to avoid being eaten.

Human Impact: Bred for home aquariums. Feeder guppies are raised as food for larger aquarium fish.

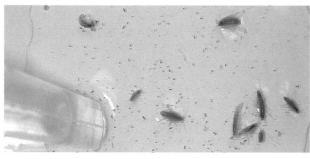

SCUD

Genus: *Gammarus*

Species: *sp.*

Size: 5–30 mm (0.2–1.2 in.) long

Range: Fresh water in the Northern Hemisphere

Natural History: *Gammarus* is much more active at night than during the day. It crawls and walks using its legs in addition to flexing its body.

Food: Bacteria, algae, and detritus (decaying organic material)

Predator: Fish, toads, salamanders, waterfowl, and crustaceans

Shelter: Detritus or plant material. Amphipods usually live close to the bottom or among submerged objects where they can hide from predators. They prefer dark areas.

Reproduction: Most scuds breed between February and October. Eggs and young develop in a brood pouch on the female. Young stay in the pouch about a week or so.

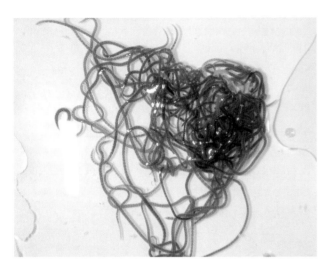

Natural History: *Tubifex* worms live on the bottom of ponds with their heads stuck into the substrate and tails waving in the water.

Food: Bacteria and detritus (decaying organic material)

Predator: Fish, amphibians, and crustaceans

Shelter: Burrows into the soil and gravel at the bottom of ponds

Human Impact: Raised as food for tropical aquarium fish

TUBIFEX WORM

Genus: ***Tubifex***

Species: ***sp.***

Size: Up to 4 cm (1.5 in.) long

Range: Temperate freshwater ponds and streams

Natural History: Snails have two sets of tentacles on their heads. The upper set contains nerve cells that are sensitive to light and smell. The two smaller tentacles on the bottom are sensitive to touch and are used to detect food, other snails, and surfaces. Snails are ectothermic and become inactive if too cool.

Food: Decaying plant material, plants, and calcium sources

Predator: Birds, skunks, and raccoons

Shelter: Snails live on the ground and will move to cool, damp places to escape dry conditions. They retreat inside their shells and form a crusty layer over the opening to preserve the moisture in their bodies. This semihibernation is called estivation.

Reproduction: Snails have both male and female sex organs (they are hermaphroditic). During mating, two snails exchange sperm; both may lay eggs. Eggs are laid underground.

Abiotic Impact: Absence of calcium in the environment causes weak shells.

Human Impact: Snails are often considered a garden and agricultural pest. For that reason, the USDA controls the movement of *Helix* across state lines. *Helix* is quarantined from some states. Some species of snail are eaten as a delicacy.

LAND SNAIL

Genus: ***Helix***

Species: ***sp.***

Size: Shell up to 3 cm (1.2 in.) diameter

Range: Worldwide

ELODEA

Genus: ***Elodea***

Species: ***canadensis***

Size: Sprigs up to 1 m (3 ft.) long

Range: Throughout North America

Natural History: *Elodea* grows in freshwater ponds and slow-moving streams throughout North America. Member of the tape-grass family.

Food: Photosynthesis

Predator: Fish, insects, aquatic snails, crayfish, turtles, and salamanders

Shelter: Freshwater ponds and quiet streams

Reproduction: *Elodea* usually reproduces vegetatively. It may produce small flowers at the tip, with seeds in small green capsules.

Abiotic Impact: Productivity depends on light, water, and temperature levels.

Human Impact: *Elodea* is a popular aquatic plant for home aquariums. Populations that have been introduced to the wild can reproduce quickly and clog natural waterways.

DUCKWEED

Genus: ***Lemna***

Species: ***minor***

Size: 0.5 cm (0.2 in.) diameter

Range: Temperate freshwater ponds and lakes worldwide

Natural History: Duckweed is tiny, but huge populations can cover the entire surface of ponds and lakes. Member of the duckweed family.

Food: Photosynthesis

Predator: Fish, aquatic snails, birds, water rats, and turtles

Shelter: Surface of freshwater ponds and lakes

Reproduction: Mostly by vegetative reproduction; sometimes sexual reproduction, if flowers appear.

Abiotic Impact: Productivity depends on light, water, and temperature levels.

Human Impact: Sold in aquarium stores for home aquariums

ALFALFA

Genus: ***Medicago***

Species: ***sativa***

Size: 30–90 cm (1–3 ft.) tall

Range: Temperate grasslands

Natural History: Alfalfa has a high tolerance for drought, cold, and heat. It has a taproot that may grow as deep as 15 m (50 ft.). Member of the pea family.

Food: Photosynthesis

Predator: Insects, birds, rodents, deer, and grazing livestock

Reproduction: Small bunches of flowers at the ends of the stems develop into coiled seed pods. Flowers are pollinated by insects, primarily solitary bees.

Abiotic Impact: Productivity depends on light, water, and temperature levels.

Human Impact: Alfalfa is cultivated as a pasture crop or for hay. Growing alfalfa improves soil quality.

RYE GRASS

Natural History: Rye grass grows in areas that are unfavorable for other cereal grains. It can be grown as a winter crop and thrives at high altitudes. Member of the grass family.

Food: Photosynthesis

Predator: Birds, insects, rodents, deer, and grazing livestock

Reproduction: Flowers are wind-pollinated.

Abiotic Impact: Productivity depends on light, water, and temperature levels.

Human Impact: Rye is an important grain cultivated throughout the world.

Genus: *Lolium*

Species: *sp.*

Size: 1–2 m (3–6 ft.) tall

Range: Temperate grasslands of Europe, Asia, and North America

WHEAT

Natural History: Wheat grows best in a temperate climate with rainfall between 30 and 90 cm (12 and 36 in.) per year. Member of the grass family.

Food: Photosynthesis

Predator: Birds, insects, rodents, deer, humans, and grazing livestock

Reproduction: Flowers are wind-pollinated.

Abiotic Impact: Productivity depends on light, water, and temperature levels.

Human Impact: Wheat is an important grain, distributed throughout the world by humans.

Genus: *Triticum*

Species: *sp.*

Size: 30–90 cm (12–36 in.) tall

Range: Grasslands worldwide

GLOSSARY

abiotic – Nonliving.

adaptation – Any trait of an organism that increases its chances of surviving and reproducing.

alkaline lake – A salty lake where the pH is greater than 7.

allele – Variations of genes that determine traits in organisms; the two corresponding alleles on two paired chromosomes constitute a gene.

aquatic – Of the water.

autotroph – Organisms that make their own food.

biomass – The total organic matter in an ecosystem.

biotic – Living organisms and products of organisms.

carbohydrate – Food in the form of sugar or starch.

carrying capacity – The maximum size of a population that can be supported by a given environment.

chromosome – A structure that transfers hereditary information to the next generation.

community – All the interacting populations in a specified area.

consumer – An organism that eats other organisms.

decomposer – An organism that consumes parts of dead organisms and transfers all the biomass into simple chemicals.

detritivore – An organism that eats detritus, breaking the organic material into smaller parts that a decomposer could use for food.

detritus – Small parts of organic material.

dominant allele – A form of a gene that is expressed as the trait when a dominant allele is present.

ecosystem – A system of interacting organisms and nonliving factors in a specified area.

environment – The surroundings of an organism including the living and nonliving factors.

exoskeleton – A tough, outer covering that insects and other organisms have for protection.

feature – A structure, characteristic, or behavior of an organism, such as eye color, fur pattern, or timing of migration.

food chain – A sequence of organisms that eat one another in an ecosystem.

food pyramid -- A kind of trophic-level diagram in the shape of a pyramid in which the largest layer at the base is the producers with the first-level, second-level, and third-level consumers in the layers above.

food web – All the feeding relationships in an ecosystem.

gene – The basic unit of heredity carried by the chromosomes; code for features of organisms.

genotype – An organism's particular combination of paired alleles.

herbivore – An organism that eats only plants.

heterotroph – An organism that cannot make its own food and must eat other organisms.

heterozygous gene – A gene composed of two different alleles (a dominant and a recessive).

homozygous gene – A gene composed of two identical alleles (e.g., both dominant).

incomplete metamorphosis – A process of gradual maturing of an insect (egg, nymphal stages or instars, adult).

individual – One single organism.

instar – An immature nymphal stage of an insect as it grows into an adult form.

limiting factor – Any biotic or abiotic component of the ecosystem that controls the size of the population.

molting – The process of shedding exoskeleton in order to grow.

morph – A form in a species that occurs in one or more forms (such as colors).

natural selection – The process by which the individuals best adapted to their environment tend to survive and pass their traits to subsequent generations.

omnivore – A consumer that eat both plants and animals.

organism – A living thing.

phenotype – The traits produced by the genotype; the expression of the genes.

photosynthesis – The process by which producers make energy-rich molecules (food) from water and carbon dioxide in the presence of light.

phytoplankton – A huge array of photosynthetic microorganisms, mostly single-celled protists, that are free-floating in water.

population – All the individuals of one kind (one species) in a specified area at one time.

proboscis – A tubelike beak for sucking fluids from plants. True bugs have this structure.

producer – An organism that is able to produce its own food through photosynthesis.

recessive allele – A form of a gene that is expressed as the trait only when a dominant allele is not present.

reproductive potential – The theoretical unlimited growth of a population over time.

species – A kind of organism; members of a species are all the same kind of organism and are different from all other kinds of organisms.

terrestrial – Of the land.

tertiary – Third level.

trait – The specific way a feature is expressed in an individual organism.

trophic levels – Functional role in a feeding relationship through which energy flows.

tufa tower – A naturally occurring, gray, lumpy structure that forms under water in a salt lake because of a chemical reaction between calcium and salt in the water.

variation – The range of expression of a trait within a population.

zooplankton – Microscopic adult animals and larval forms of animals found free-floating in fresh water and seawater.